Solar System Photometry Handbook

FRAMINGHAM STATE COLLEGE
3 3014 00086 1395

U0880190

Copyright © 1983 by Willmann-Bell, Inc.
All Rights Reserved

Reproduction or translation of any part of this work beyond that permitted by Sections 107 or 108 of the 1976 United States Copyright Act without permission of the copyright owner is unlawful. Requests for permission or further information should be addressed to the Permissions Department, Willmann-Bell, Inc., P.O. Box 3125, Richmond, Virginia 23235 USA.

Library of Congress Cataloging in Publication Data

Main entry under title:
Solar system photometry handbook.

Includes bibliographical references and index.
1. Photometry, Astronomical. 2. Photoelectric measurements. I. Genet, Russell.
QB 135.S68 1983 522′.623 83-21382
ISBN 0-943396-03-4

Printed in the United States of America
10 9 8 7 6 5 4 3 2 1

Designed by Erdvilas K. Bankauskas

Solar System Photometry Handbook

Edited by Russell M. Genet

Assistant to the Editor:
Dorothy K. Mote

Forewords by Douglas S. Hall
and David W. Dunham

Authors:

Richard P. Binzel	Photometry of Asteroids
G. Wesley Lockwood ...	Photometry of Planets and Satellites
Michael F. A'Hearn	Photometry of Comets
Peter Hedervari	Lunar Photometry
Gary A. Chapman	Solar Photometry
Jeffrey L. Hopkins	Low Speed Equipment
Robert L. Millis	Occultations by Planets and Satellites
Alan W. Harris	Asteroid Occultations
Graham L. Blow	Lunar Occultations
Peter C. Chen	Portable High Speed Photometer Project

Framingham State College
Framingham, Massachusetts

Published by:

Willmann-Bell, Inc.

P. O. Box 3125
Richmond, Virginia 23235 ☎ (804)
United States of America 320-7016

About the Authors. . .

RUSSELL M. GENET received his B.S. in Electrical Engineering from the University of Oklahoma in 1964 and a M.S. from the Air Force Institute of Technology. He is Chief of Acquisitions Logistics Research at the Air Force Human Resources Laboratory. He has authored many papers and reports in engineering and operations research. He is author of "Real-Time Control with the TRS-80", and co-author of "Photoelectric Photometry of Variable Stars". In 1979, he founded the Fairborn Observatory which is devoted to photoelectric observations of eclipsing binary stars. He is co-editor of the IAPPP Communications.

RICHARD P. BINZEL received his BA in Physics and Mathematics (With Highest Honors) from Macalester College in 1980. In 1982 the University of Texas at Austin awarded him a MA in Astronomy. In the past he has worked at the U.S. Naval Observatory, California Institute of Technology, and the University of Arizona Steward Observatory. He is currently a candidate for the Ph.D. degree at the University of Texas where he is also a Research Assistant on the Pluto Project. He has authored or co-authored more than 20 scientific papers and now is editor of the ALPO Minor Planet Bulletin.

PETER HEDERVARI studied astronomy and geo-sciences at the Roland Eotvos University, Budapest, Hungary which awarded him a Ph.D. His thesis was on planetary volcanism. He has authored more than a dozen books and over 200 scientific papers on volcanology, seismology, plate tectonics and planetology. He has been active in numerous international scientific societies. Currently, he is associated with the Georgiana Observatory at Budapest where he conducts hydrogen alpha observations of limb prominences and flares. The data from these observations are published by the World Data Center for Solar-Terrestrial Physics at Boulder, Colorado.

ROBERT L. MILLIS was awarded a B.A.in physics (With High Honors) by Eastern Illinois University in 1963. In 1968, the University of Wisconsin at Madison awarded him a Ph.D. in astronomy. He is currently associated with the Lowell Observatory, Flagstaff, Arizona where he is both a Staff Astronomer and Secretary-Treasurer. Additionally, he is Adjunct Professor of Astronomy at Ohio State University, Columbus. His research experience includes investigation of asteroids, ring systems, planetary atmospheres through observation of stellar occultations, spectrophotometry of comets, photometry of planetary satellites using area-scanning and conventional techniques.

MICHAEL F. A'HEARN was awarded a B.S. in Physics by Boston College in 1961. The University of Wisconsin at Madison awarded him a Ph.D. in Astronomy in 1966. Since 1966 he has been associated with the University of Maryland where he is now Professor of Astronomy. He is a member of numerous professional societies and has written many papers on comets. He is Discipline Specialist for Photometry and Polarimetry for the International Halley Watch. He is also Chairman for IAU Commission 15, Working Group on Standardized comet filters.

G. WESLY LOCKWOOD was awarded a B.S. in Physics from Duke University in 1963. He did his graduate work in astronomy at the University of Virginia, earning a M.S. (1965) and Ph.D. (1968). He is an astronomer at Lowell Observatory, where he specializes in highly accurate "milli-mag" photoelectric photometry. He is widely known for his measurements of the brightness of planets to determine possible solar variability. Prior to coming to Lowell Observatory he was employed as an astronomer at Kitt Peak National Observatory.

PETER C. CHEN earned his Ph.D. in Astronomy at Case Western Reserve University. He did post doctorial research at The University of Texas at Austin where he worked with Dr. R. Edward Nather, the "Father" of high-speed photometry. He is currently working at NASA's Goddard Space Flight Center where he is employed by a contractor providing computer software.

QB 135 S68 1983

GARY A. CHAPMAN received his Ph.D. in 1968 in astronomy at the University of Arizona, Steward Observatory. He has worked on the staff of the Institute for Astronomy at the University of Hawaii and the Aerospace Corporation where he did research in solar physics. While with the Aerospace Corporation he worked on the x-ray telescope experiment that flew with Skylab. He is currently director of the San Fernando Observatory.

GRAHAM L. BLOW received his B.S. in Physics (1976) and M.Sc. (With Honors) (1980) from the University of Auckland. He is Scientific Officer at the Carter Observatory (National Observatory of New Zealand) in Wellington. Starting as a youth he has specialized in lunar occultations. He continues to be active in numerous amateur societies and publications in New Zealand and Australia.

JEFFREY L. HOPKINS, was graduated from the University of Syracuse in 1966 with a B.S. in Physics. He has also done graduate work at the University of Wyoming. He is currently a Senior Staff Engineer with Motorola. He has designed the "HPO" line of photoelectric photometry equipment. He also has written numerous papers and is the Editor of the Epsilon Aurigea Newsletter.

ALAN W. HARRIS was graduated for California Institute of Technology in 1966 with a B.S. in Geophysics. His graduate work was in Planetary and Space Science at the University of California at Los Angeles where he earned his M.S. (1967) and PH.D. (1975). He worked at Rockwell International, taught at the NATO Advanced Study Institute and been a guest investigator at the Hale Observatories. He was a Visiting Associate Professor at both the University of California at Santa Barbara (Physics) and University of California at Los Angeles (Earth and Space Science). Since 1974, he has been a Member of the Technical Staff of Jet Propulsion Laboratory.

TABLE OF CONTENTS

SECTION B: HIGH SPEED PHOTOMETRY

PREFACE

This book is a result of and testimony to the cooperation that exists between large and small observatories--between professional and amateur astronomers. The book is dedicated to the seemingly contradictory notion that serious scientific observations of solar-system objects can be made with the small telescopes of backyard amateurs and small-college professors and students. Astronomy remains one of the few sciences where, with a modest investment of equipment and time, the serious amateur can make important and solid contributions to the advancement of scientific knowledge. Most professional astronomers welcome such amateur contributions and openly assist them. There is little, if any, bias against amateurs publishing their results in the leading astronomical journals. Thus, for the serious amateur scientist, astronomical research based on backyard observations remains an important haven. The same attractions hold for the small-college professor and his students. The ability of an undergraduate to do a complete piece of astronomical research as a senior project, including the gathering and analysis of the data, and having this research published in a recognized journal, does much for the advancement of the scientific spirit in education at smaller colleges.

A large portion of the serious astronomical science accomplished at smaller observatories is based on photoelectric observations. The reason for this is that wide-band photometry makes very efficient use of the meager photons available to smaller telescopes. Photometry is also surprisingly tolerant of bright background lights, a frequent occurrence in the suburbs and small cities where most amateur and small college observatories are located.

It is the intent of this book to provide a concise introduction to solar system photoelectric photometry as currently practiced at smaller observatories. Increasing numbers of advanced amateur and small college observatories are equipping themselves with the relatively low cost and now readily available equipment required to make photoelectric observations. While the majority of these smaller observatories have chosen to concentrate initially on the more traditional stellar photometry--especially variable star photometry--many small-observatory photometrists have expressed an interest in initiating observations of selected solar system objects within their capabilities. There are also many smaller observatories where observations of solar-system objects has always been the primary program, but such observations have been entirely visual. A number of these observatories have indicated a desire to complement their visual observations with photoelectric observations where the latter are within their capabilities.

The advanced amateur and small college professor (and student) have four important advantages over their large-observatory brethren that can have an important influence on the future of solar system photometry. The first advantage is the sheer number of small-observatory astronomers. The number of astronomers at the larger observatories and universities who devote most of their time to observations of solar system objects is very small, and the number doing photometry of these objects on a regular basis is the merest handful. This is not due to any lack of important scientifically interesting projects that could be undertaken, but due simply to a lack of astronomers. Many important, easily made observations go begging. On the other hand, the number of smaller observatories manned by part-time astronomers making uninstrumented observations of solar system objects is very large indeed. While such visual observations continue to be of some value, photoelectric observations are much more precise and hence much more valuable.

The second advantage of the small observatory proprietor is the availability of telescope time. The larger observatories dole out their telescope time in small blocks that are scheduled many months in advance. For many observational programs in solar system photometry, the observations must be made at a precise point in time, and no other will do. Other observational programs require long periods of time on the telescope.

The third advantage of the smaller observatory is the diversity of location. Rather than being concentrated at a few locations of particularly good weather and seeing, smaller observatories are spread all about. Many solar system observations can be made only from a specific location on the surface of the Earth. In many cases this means that, if the observations are to be made at all, they _must_ be made by smaller observatories.

Finally, small observatories tend to have small telescopes, and small telescopes tend to be portable. For some observations, such as asteroid occultations, a large number of observers need to be placed at rather precise locations in a pattern. This can be done only with portable telescopes--a common device at smaller observatories, but something of a rarity at larger observatories.

The four advantages cited above give small observatories the _potential_ for making many significant contributions to the advancement of solar-system astronomy. This potential remains largely untapped for three reasons.

First, very few smaller observatories know about the need for photoelectric observations of solar system objects. Where this need is known, it is thought not to be within the capabilities of smaller observatories. Hopefully, this book will help to reaffirm the need and dispel the notion that such observations are not within the capabilities of smaller observatories.

Second, until recently, smaller observatories desiring to make photoelectric observations had to build their own photometers. It is really too much to expect professors to work or teach all day, make observations at night, _and_ build their own equipment! However, in just the last year or two, several excellent commercially-made photometers have become available at very reasonable prices. The small observatory photometrist no longer needs to make his own equipment.

Finally, photoelectric occultation timings require high-speed recording equipment that, until recently, required complex and expensive equipment--placing such observations beyond the reach of most smaller observatories. The University of Texas, a long-time leader in high-speed photometry, recognized this difficulty and developed a relatively low-cost recording system. This key development is reported by Dr. Chen in this book and it is hoped that, shortly after this book appears, this recording device will be commercially available at a price within reach of smaller observatories.

Thus, while very good reasons existed in the past for the considerable potential of smaller-observatory photometry not being realized in practice, these barriers have now been removed and a lively growth can be expected in this area over the next several years. It is hoped that this book will contribute to this growth and the tapping of this rich potential.

Compared to variable star photometry, solar system photometry is somewhat different, sometimes considerably so. The "slow speed" solar system photometry is most similar to variable star photometry so it has been grouped together in the first section of this book. Photometry of asteroids is almost identical to the photometry of short-period variable stars. The only major difference is that the asteroids change their position from night-to-night. Thus, one needs to be somewhat adept at locating the asteroids and occasionally choosing new comparison stars. There are, of course, a few fine points involved, and Richard P. Binzel, Department of Astronomy, University of Texas, tells us about these in Chapter 1. As Chairman of the IAPPP Asteroid Committee, Editor of the Minor Planet Bulletin (ALPO), and an active photometrist on telescopes large and small, he is well-qualified to cover this topic.

Useful planetary photometry is challenging because very high precision is required. Photometry of planetary satellites has the added challenge of measuring faint objects close to a bright one, and one needs to worry about gradients in the backgrounds, among other things. G. Wesley Lockwood is a leading expert on precision planetary photometry, and he provides the details the small observatory photometrist needs to get started in Chapter 2. Not surprisingly, he does his photometry at Lowell Observatory in Flagstaff, Arizona--an observatory long noted for its planetary research program.

While still "slow speed," cometary photometry has the interesting complications of photometry in twilight near the horizon. Special narrow-band filters are used to study particular portions of the spectrum. Michael F. A'Hearn at the University of Maryland is the widely recognized expert in this area of photometry. He is the Chairman of the IAPPP Cometary Photometry Committee, and is the

Discipline Specialist for Photometry in the International Halley Watch (IHW) program. Read Chapter 3 and get ready for Halley's comet!

Somewhat brighter than comets is our own nearby Moon. The first photometry of the Moon was by Joel Stebbins in 1907, and surprisingly little has been done since then. Lunar photometry is carefully explained by Peter Hedavari, Director of the Georgiana Observatory in Budapest, Hungary.

Brightest of all, of course, is the Sun. Photometry of the Sun has two distinct advantages. First, it can be done during the day when normal people are awake. Second, generally there is no shortage of photons to count. A large telescope is not needed for solar photometry. Solar photometry is a large field, and one chapter can not do it justice, but Gary A. Chapman, California State University of Northridge, nicely introduces several aspects that are particularly suited to small observatories in Chapter 5.

While slow-speed photometry equipment is covered in several recent books to some extent, it is a rapidly changing area, and Jeffery L. Hopkins of the Hopkins Phoenix Observatory kindly provides us with an updated summary in Chapter 6. His "HPO" series of photoelectric equipment is widely known for both its small size, modest cost, and being the only photometer in the world available as a kit as well as a fully assembled unit.

The second section of the book is devoted to "high speed" photometry. Perhaps photometry of planetary occultations and asteroid occultations might more appropriately be termed "medium speed" photometry similar to flare star photometry. However, there is no question that lunar occultation photometry is "high speed."

The planets occasionally occult a star or their own satellites, and comets must eventually occult a star. These events are exciting and "fast paced." Often they can only be seen from limited areas of the Earth's surface, and a smaller observatory in the right location that is equipped for photometry can provide a very valuable and unique service by observing such events. Robert L. Millis provides the details in Chapter 7. He is an astronomer at Lowell Observatory, and an expert on many aspects of solar system photometry. That two astronomers from Lowell Observatory should be included in this book is entirely appropriate.

Asteroid occultations are also a fast-paced and exciting game. Properly approached, asteroid photometry requires mobility on very short notice, and a large number of observers working as a team. Small portable telescopes are the "stock in trade" of the small observatories, and thus they are ideally suited to carry on this important scientific work. Alan W. Harris of the Jet Propulsion Laboratory (JPL) is an active photoelectric asteroid occultation observer, and he spells out just what is required in Chapter 8.

Our Moon, being close to us, appears to move more rapidly against the background of stars, and as a result, lunar occultations happen very rapidly. It is all over in less than a second. Without question, this is HIGH speed photometry. R. Edward Nather and David S. Evans, both of the University of Texas, pioneered high speed photometry and its application to lunar occultations. They have passed on their techniques and knowledge to others, including Graham Blow, the scientific officer at Carter Observatory--the national observatory of New Zealand. In Chapter 9, Graham Blow explains how lunar occultations are observed and analyzed.

As might be expected, special recording equipment is needed for lunar occultations, and in Chapter 10, Peter C. Chen, appropriately from the University of Texas, describes a relatively low-cost occultation recorder. There are plans to manufacture this device commercially, and if this happens it will revolutionize the observations of lunar occultations at smaller observatories.

My role as editor has been that of defining the areas to be covered by the book, selecting experts who were willing to write the various chapters, editing the manuscript, and arranging for its publication. The bulk of the credit for this book goes, of course, to the authors of the individual chapters, and to Willmann-Bell, Inc., the publisher of this book. It speaks well of the health of astronomical science, that eminent astronomers at some of the largest observatories are not only willing to take time to write chapters for a book intended for use at the smallest observatories, but are vocal advocates for increased scientific work amongst backyard amateur astronomers.

It is my intent that in a few years this book will be reissued in a revised edition. Suggestions for improving the book, correction of errors, and ideas for added chapters or material would all be welcome. Please send these to me at: Fairborn Observatory, 1247 Folk Road, Fairborn, Ohio 45324 USA. All comments will be routed to the appropriate author, and will also be retained for inclusion in the revised edition as appropriate.

My special thanks is due to Dorothy K. Mote, the "producer" of the book. She not only typed the manuscript, but arranged the figures and provided the artistic touches that I would never have thought of. My thanks also to Robert Thomas, whose interest in planetary photometry helped to instigate this book. We planned to edit this book together, and it was only his pressing work as a geologist and an untimely illness that prevented this from happening.

Finally, my thanks and appreciation to the center of my little solar system--my wife Ann. She continues to encourage me in my various efforts, including the editing of books--a real time consumer. May she never be eclipsed or occulted!

Russell M. Genet
Fairborn Observatory

FOREWORD

How nice of my coauthor on another book to ask me to write a foreword to this book. Let me explain why this was so appropriate.

When Russ Genet and I were conceiving "Photoelectric Photometry of Variable Stars," we wrestled with a big question: What should be the scope and, consequently, the title? It was clear we wanted to share with the world our conviction that the technique of photoelectric photometry can produce really valuable scientific results, not only at a major professional installation but also at a small campus college observatory or a backyard amateur observatory. The task was not easy because so much ground needed to be covered. Our treatment of the relevant equipment, instrumentation, electronics, and observing techniques (which we hope was adequate) alone could have made one book. But it was mandatory, we felt, to indicate in specific detail what could be done with this powerful photometric technique. The world's best observatory, large or small, will be wasted totally if it is not used.

So much can be done with photoelectric photometry. The long list of potential targets includes: variable stars, star clusters, tumbling asteroids, comets, occultations by the moon or by planets and asteroids, comparison star sequences, galaxies, and the sky itself, for light pollution studies. Largely at my insistence, we decided to focus on just one target: variable stars. Another reason was that those bedeviling variable stars had for years been my own research specialty and I felt comfortable with them. The result was "Variable Stars" in the title. But I never denied or forgot that photoelectric photometry can do much more than variable stars, and have continued to feel a sin of omission was committed.

It is ironic that just recently I have become involved in some exciting solar system research myself, at least indirectly. One of my graduate students, Robert L. Marcialis, is attempting to understand the light variations of Pluto in terms of a computer model in which the surface is covered unevenly with light and dark areas. You may recall that it was Robert H. Hardie, also at Vanderbilt University's Dyer Observatory, whose photoelectric photometry led to the original discovery of Pluto's 6^d9^h rotation period. If Marcialis succeeds, the result will be the first map of the surface of Pluto, the unresolved planet! Thus, my own example is a good example of the wonders photoelectric photometry can accomplish.

This new book, covering virtually every aspect of solar system photometry, carries us much farther towards the goal Russ and I originally had. Seeing it go to press now, I feel absolved of my sin of omission. It would please me immensely to have "Solar System Photoelectric Photometry" considered a companion to "Photoelectric Photometry of Variable Stars."

Perhaps these two books will be joined subsequently by other companions, perhaps not. Whatever happens, I invite all interested in photoelectric photometry, amateurs and professionals, those with telescopes large and small, to participate actively in International Amateur-Professional Photoelectric Photometry. A reader of the quarterly I.A.P.P.P. Communications will continue to see timely articles dealing with all aspects of photoelectric photometry, from transistors to galaxies.

Douglas S. Hall
I.A.P.P.P.

FOREWORD

To date, the high-speed automatic recording of rapid Solar System phenomena has been almost exclusively the domain of a few groups of professional astronomers at major observatories who have received grant support to develop sophisticated equipment and make observations. Although the results of their observational programs have been impressive, all agree that much more valuable scientific data could be obtained if high-speed recording systems were more widely distributed and used. Very few amateur astronomers and small-college observatories have joined the game, primarily because those with the electronic skills to build state-of-the art high-speed recording systems are often not very interested in observing, and vice versa.

The second section of this book should solve this program. The chapters by Millis, Blow, and Harris lay the theoretical groundwork by describing the nature of the rapid light variations which occur during occultations and eclipses, and the astronomical and astrophysical information which can be obtained through analysis of photometric observations of these phenomena. In addition, Chen describes a recording system which observers with electronic skills can assemble. Arrangements are now being made to manufacture the University of Texas' system designed by Chen.

In the paragraphs below, I first discuss the value of Lunar occultation timings and the sources of errors contributing to their analysis. I explain why visual occultation timings are still needed, as well as the advantages of photoelectric observations. The relative values of the two types of observation for asteroidal and planetary occultations are also briefly assessed. I conclude with information about obtaining occultation predictions, especially those provided by the International Occultation Timing Association (IOTA).

Value of Visual and Photoelectric Occultation Observations

Lunar occultations have been timed visually for over 300 years. The observations have been analyzed to determine improved mean orbital elements of the Moon; corrections to the system of right ascension and declination (the equinox of the new FK5 fundamental celestial coordinate system has been determined largely from occultation observations); individual star positions and proper motions; discovery of close double stars which cannot be resolved by direct visual or spectroscopic techniques; corrections to the reference system, or "datum," for the extensive Lunar profile limb corrections compiled by C. B. Watts; variations in the rotation of the Earth (determination of ΔT, the difference between Ephemeris and Universal Times); and the secular acceleration of the Lunar longitude which depends on tidal interactions and possible rate of change of G, the universal gravitational constant, postulated by many cosmoligical theories. Equinox and obliquity corrections determined from occultations have even been used to derive parameters of Galactic rotation, since these are related through the system of stellar proper motions. The improvement of Watt's limb data from occultations leads to an improvement of the analysis of total and annular Solar eclipse contact timings, which have shown small variations of the Solar radius during the last two centuries.

Visual timings are still useful for most of the above applications, and should not be discouraged by increasing numbers of photoelectric observations. For example, a visual timing accurate to $\pm^{s}.2$ defines s, the position of the star relative to the Lunar limb, to less than $\pm 0''.1$. During grazing occultations, the geometry helps so that accuracies of a few hundredths of an arcsecond can be achieved with timings accurate to only a second. In 1971, L. V. Morrison published "A Comparative Study of Visual and Photoelectric Timing of Occultations" in *Highlights of Astronomy* (De Jager, ed., D. Reidel Pub. Co.). He estimated errors contributing to s from timing, star place, the Lunar ephemeris, and Watts' limb profile, obtaining $0''.26$, $0''.31$, and $0''.20$, respectively, for the last three. These are much larger than the error for a good visual timing mentioned above, so that reducing the timing error to essentially zero by photoelectric timing does little to improve the overall analysis. Morrison estimated the accuracy of an average visual timing from an analysis of about 1000 visual timings of occultations of 22 Pleiades stars on 1969 March 23, and obtained $0''.21$, corresponding to a timing accuracy of $\pm 0^{s}.44$. He noted that this was larger than expected, because large numbers of new (inexperienced) observers may have augmented the value. Another analysis of a larger set of timings implied a visual timing accuracy of about $0^{s}.27$. Morrison concluded that, due to the dominance of the other sources, photoelectric observations had only slightly greater weight for occultation analyses than visual timings, in a ratio of about 6 to 5.

Since Morrison published his study, the Lunar ephemeris error has been virtually eliminated through better theoretical models and analysis of laser rangings to the retroreflector arrays placed on the Moon. The U.S. Naval Observatory's improved XZ-catalog data for the northern Zodiacal stars and Perth 70 data for the brighter southern stars have considerably reduced star place errors. Star place errors may be reduced to insignificance during the next 10 to 20 years from astrometric spacecraft and Earth-based interferometric observations. But little has been accomplished to reduce the Lunar limb error, and I know of no plans to do this in the future, except via occultation timings, especially those made during total Lunar eclipses. Consequently, good visual timings are still very useful, especially of reappearances, which are difficult to record photoelectrically. Morrison's photoelectric/visual ratio is probably now nearly 2 to 1, and will probably increase very slowly during the next twenty years. Since most of the remaining error is in the Lunar profile, visual observations of grazing occultations still have essentially equal weight with photoelectric observations. Since grazes must usually be observed with portable equipment in the field, only a couple of them have been recorded photoelectrically, so visual observers will continue to have the most impact for Lunar occultation analysis in this area. The grazing geometry permits closer double stars to be resolved, and the main drop of light due to diffraction is often evident visually as a "gradual" disappearance or reappearance.

If photoelectric timings give only a 2-to-1 advantage over visual timings (and not even that for grazes), you may ask why you should go to the effort of buying or constructing a photoelectric system? One reason is that the discussion of errors above assume random errors; if there is an unknown systematic error in the visual timings, for example, in the estimation of personal equations, some of the parameters derived from the analysis will have corresponding unknown errors. This is especially important for the acceleration of the Moon's longitude and consequent determination of rate of change of G; hence, photoelectric timings are used almost exclusively for this work. Systematic errors in visual timings can be studied by visual and photoelectric timings of the same occultation at the same place. Such observations are strongly encouraged, since only visual timings are available, with only a few exceptions, before 1950. Quite a few simultaneous visual-photoelectric timings have been made, mostly in Japan, but more are needed for all types of visual timings, including eye-and-ear, stopwatch, and tape-recorder or chronograph.

The main value of a photoelectric observation is that it provides a record of the variation of the star's light with time. In the case of close double stars, this permits the determination of the relative brightness and accurate relative separation (in the direction perpendicular to the Lunar limb) of the components. A visually-observed "gradual" occultation could be due to either diffraction or close stellar duplicity; the cause would be unambiguous with a photoelectric record. Very close (of the order of a few milliarcseconds) stellar duplicity and stellar angular diameters can only be determined from occultations by high-speed photoelectric photometry. A record of the variation of the star's light is essential in the case of occultations of stars by planets with atmospheres, the visual observation of which is almost useless, even for bright stars.

The permanent record obtained from a photoelectric observation also gives an advantage over visual timings. A given photoelectric record may seem too noisy to tell if a star is double or not. A later observation, or one made elsewhere the same night, might clearly show duplicity. The record can then be examined to locate the two stars and measure their separation in a different position angle. Since observations of the same occultation from two or more widely-separated observatories are needed to determine both the separation and position angle of a close double, more photoelectric observers are needed to increase the possibility that at least two of them will have clear skies for a given occultation. An even stronger argument exists in the case of secondary occultations of stars by asteroids or planets which could provide indisputable evidence for a new satellite.

The angular motions of Solar System objects beyond the Moon are so small that any occultation observation, either visual or photoelectric (and in some cases, even if a timing is not made), will result in a substantial improvement of the position of the object relative to the star. Future astrometric observations of the star, such as by ESA's planned HIPPARCOS satellite, can then be made to significantly improve the orbital elements of the occulting body (three or more occultations by the same body are needed to do this accurately; otherwise, less-accurate photographic positions are also needed.

For asteroidal occultations, the timings should be accurate to within about 2 percent of the central occultation duration in order to determine an accurate mean diameter and shape of the asteroid. Hence, carefully-made visual timings are quite useful for occultations with predicted central durations greater than 10 seconds, but are of not much use for this purpose if the central duration is 5 seconds or less. Asteroids with diameters less than 100 km can occult a star for over 10 seconds near a stationary point, but the asteroids larger than 250 km produce occultations longer than 10 seconds more frequently. These larger asteroids are often brighter than the occulted star, rendering visual observation very difficult, if not impossible. Hence, a significant fraction of predicted asteroidal occultations cannot be observed visually, and this fraction will probably increase as more fainter stars are considered for predictions. Since more observed occultation chords define the size and shape of the asteroid more accurately, the largest number of potential observers possible is needed for each event. Many events can either be detected only photoelectrically, or are too short to time accurately enough visually. More photoelectric observers also increases the chances of obtaining unambiguous and/or confirmed records of secondary occultations. Since asteroidal occultation paths are narrow and can often be predicted to less than their width via last-minute astrometry, portable equipment (either for field use or transportable to, and usable at, another observatory) is highly desirable.

Predictions and I.O.T.A.

Approximate information about the most favorable Lunar occultations, generally for stars brighter than 5th magnitude, is published each year in my Occultation Highlights article in the Observer's Page section of the January issue of Sky and Telescope. Although the emphasis is on events visible from North America, some information about other interesting Lunar occultations, such as those of major and minor planets, star clusters, and nebulae, is included for events visible from Europe and other parts of the world as well. Sources for more detailed predictions and services of the International Occultation Timing Association (I.O.T.A.) are also given.

The Observer's Handbook of the Royal Astronomical Society of Canada includes an Occultation Supplement which contains predictions for 18 American and Canadian "standard stations," listing about 50 events of stars to 6th magnitude for each station and including linear factors which can be used to compute approximate times for locations within about 200 miles of each station. It also contains maps showing the northern and southern limits of over 100 occultations; favorable grazing occultations are visible within about two miles of these lines. An offprint of the Occultation Supplement can be obtained for 50 cents and a long, self-addressed unstamped envelope from the Royal Astronomical Society of Canada, 124 Merton Street, Toronto, Ont. M4S 2A2. Similar material is available in the annual astronomical publications of other nations. These published predictions are prepared by the International Lunar Occultation Centre (ILOC), Astronomical Division, Hydrographic Dept., Tsukiji-5, Chuo-Ku, Tokyo, 104 Japan. ILOC also distributes standard observation report forms and collects all Lunar occultation observations for analysis.

More extensive predictions, including stars to 8th mag. under favorable conditions, can be obtained by sending accurate coordinates and a long self-addressed stamped envelope to Walter V. Morgan, 10961 Morgan Territory Road, Livermore, California 94550; the computer program he uses applies linear factors, like those in the Occultation Supplement mentioned above, to calculate the times at local stations, using data for one of over 20 North American standard stations. Similar predictions for Europe are available from Hans Bode, Bartold-Knaust-Strasse 8, D-3000 Hannover, West Germany.

If you plan to observe Lunar occultations regularly with high-speed photoelectric equipment, you should send accurate coordinates and information about the telescope to Marie R. Lukac, Nautical Almanac Office, U. S. Naval Observatory, Washington, D.C. 20390 to obtain detailed total occultation predictions computed accurately for your location. These predictions consider all stars in the ACK3 and SAO catalogs, so there are typically several occultations listed each night. A full list for a given year usually is over 60 pages long with nearly 3000 events.

For those with portable equipment, detailed computer predictions (lists of geographical coordinates in the northern or southern occultation limit and predicted Lunar profile) of grazing occultations are available for $1.50 each from IOTA, P. O. Box 596, Tinley Park, Ill. 60477. Papers describing the use of these graze predictions can be obtained for $2.50. Alternately, you can join IOTA for $11, or $16 if you do not live in either the U.S.A., Canada, or Mexico. Membership benefits include free graze predictions (computed automatically for specified distances from your station and including stars brighter than mag. 8.6 under favorable conditions; each year, within about 100 miles of any given location, there are about two dozen grazes potentially visible with a 6-inch telescope), local circumstances of asteroid appulses and planetary occultations, the papers describing use of the graze predictions, and a subscription to Occultation Newsletter (also available separately for $5.50, plus the differential cost of airmail postage for those outside North America). The quarterly Occultation Newsletter includes detailed predictions of all known observable planetary and asteroidal occultations (with small world maps, more detailed regional maps showing possible parallel paths for each reference when last-minute predictions become available, and finder charts); small world maps showing the regions of visibility of Lunar occultations of the planets and first four asteroids, tallies of Lunar occultation timings; reports and lists of observed grazes, double stars discovered during occultations, and observed planetary and asteroidal occultations; and articles about special events (such as eclipses), observing techniques, and equipment. IOTA's services are available for all parts of the world except Europe; Europeans can obtain similar predictions and Occultation Newsletter from the European Section of IOTA headed by Hans Bode, whose address is given above. Membership in IOTA/ES costs DM 12.

Predictions of approximately the 50 best planetary and asteroidal occultations for a given year are listed in a table in my annual article about these events in the Celestial Calendar section of the January issue of Sky and Telescope; it also includes a rundown on the previous year's observations and a map of the predicted paths of events visible from the U.S.A. and southern Canada. More detailed information is given in Occultation Newsletter, as described above, and also sometimes distributed by Gordon Taylor, Royal Greenwich Observatory, England, and by astronomers at Lowell Observatory and the Massachusetts Institute of Technology. Predictions of local circumstances (time and angular separation at closest approach, altitudes and azimuths of the star, Sun, and Moon) for all planetary and asteroidal occultations and close appulses that will occur above your horizon can be obtained by sending your coordinates and $1 (checks payable to IOTA, but free for IOTA members) to Joseph E. Carroll, 4216 Queen's Way, Minnetonka, Minn. 55343. A self-addressed envelope will expedite requests. Last-minute astrometric updates of asteroidal occultation paths can be obtained from a few days to a few hours before most North American events from recorded telephone messages at 312-259-2376 (Chicago, Ill.). Europeans can obtain similar information by telephoning Taylor at the Royal Greenwich Observatory, England, 323-833171.

A final note of caution: If you construct a high-speed photoelectric system only expecting to use it for asteroidal occultations, you will probably be disappointed unless you are able to transport it over long distances. These events for a given region are quite infrequent, especially when adverse weather and astrometric errors are considered. For example, only five asteroidal occultations have been observed from the eastern U.S.A. to date, although reasonably comprehensive predictions became available only six years ago. The statistics should improve as more stars and more asteroids are considered during future searches for occultations. But you should also try Lunar occultations in order to get full use of your equipment and justify its construction and expense. My discussions above and Graham Blow's chapter should provide sufficient motivation for undertaking Lunar occultation observations.

David W. Dunham
I.O.T.A.

SECTION 1

LOW SPEED PHOTOMETRY

1. PHOTOMETRY OF ASTEROIDS

Richard P. Binzel

I. INTRODUCTION

Asteroids, or minor planets, are the periodic variable "stars" of the solar system. Their variability was not detected until a full century after the first asteroid was discovered when in 1901 E. Oppolzer noted a periodic variability in the brightness of the asteroid 433 Eros. (The number preceding the name of an asteroid roughly indicates the order in which it was discovered and is a convenient way of cataloging.)

The variability of most asteroids is attributed to their irregular shape. As an asteroid rotates about its axis, it reflects differing amounts of sunlight towards the Earth and, therefore, produces observable brightness variations. Measurement of these variations reveal a lightcurve which repeats with a period which corresponds to the synodic rotation period of the asteroid. Since the sides of an asteroid are not necessarily symmetric, a typical asteroid lightcurve displays two usually unequal maxima and minima as each of the four sides are viewed in turn. In some cases, lightcurves may be due to eclipsing or occultation events caused by objects in orbit about an asteroid. Only for 4 Vesta is the lightcurve known to be primarily due to the color variations rather than the shape.

Most asteroids have rotational periods in the range of 4 to 12 hours and lightcurve amplitudes of a few tenths of a magnitude; however, some highly irregular asteroids display amplitudes greater than 1 magnitude. The short periods and moderate amplitudes of the asteroids make them rewarding objects to observe because a good portion of a complete lightcurve can usually be observed during a single night. A single night of observation to obtain at least part of an asteroid lightcurve makes a very good exercise, although the most scientifically valuable observations require lightcurves obtained over several nights.

Color indices, usually B-V and U-B, have been observed for many asteroids. It has been found that asteroids are not the same color, but actually have colors which fall into several categories. These color differences represent distinct compositional differences in the surfaces of asteroids and most likely their interiors as well.

Photometric observations and other studies of asteroids have greatly increased in recent years as interest in the asteroid belt has grown. This growth has resulted because it is believed that an understanding of the asteroid belt can give important insight into the origin and evolution of the entire solar system. Most astronomers believe that asteroids are remnant material from the formation of the solar system, although the possibility that they are the remnants of a planet which once occupied that position and subsequently broke up can not be ruled out. Rotational periods and amplitudes obtained from lightcurve observations allow models of the collisional evolution of the asteroid belt to be constructed which may determine which of these theories is most likely correct.

II. OBSERVING PROGRAMS

Although most of the brighter asteroids have had their periods of rotation and colors measured, amateurs with photoelectric equipment can still make a valuable contribution to their study as the work on these objects is by no means finished. The number of professional astronomers making photometric measurements of asteroids is still relatively small so only a small amount of professional telescope time is devoted to asteroid study. Even when asteroids are observed by professionals, work is usually done on the fainter asteroids which can only be observed by the larger telescopes. This leaves a large number of bright asteroids whose study is left to the amateur.

Potential observing programs are listed below in two categories, exercises and programs. The exercises involve a single night of observation which demonstrate some aspect of asteroid study. The programs are listed in increasing order of scientific importance and amount of work required, the results of which may be published. Any photometric instrumentation can be used which can achieve the accuracies of 0.01 to 0.05 magnitude for individual measurements. A 1P21 with standard filters is highly desirable for programs which require transformation to the standard system. Since most of the programs utilize differential measures, observations can be made under less than perfect photometric conditions providing comparison star observations can be made quite frequently. Good conditions are required for color observations and transformation to the standard system. No filter is necessary for differential observations which will not be transformed.

As seen from the following descriptions the most scientifically useful and rewarding programs involve the selection of a single asteroid which is observed repeatedly over many nights over a long period of time. This may be accomplished by one observer or more easily by a group of observers. The philosophy here is that it is better to have one asteroid well observed than to have scattered observations of many different asteroids.

Observing procedures for the programs listed are described in Section IV and reduction procedures in Section V.

Exercises

1. Observe a lightcurve. Obtain a lightcurve by differentially observing an asteroid known to have a moderate amplitude over as many hours as possible on a single night. Any or no filter may be used. If all or most of the rotational period is observed, these observations may be valuable when combined with those of other observers. In this case the observations are worth reporting.

2. Observe color differences. Measure the instrumental colors of several different asteroids on the same night. If possible choose asteroids of varied compositional types so that color differences will be more easily observable. Compare results of instrumental B-V and U-B color indices to see the color differences among asteroids. These observations need not be transformed to the standard system. This exercise requires good photometric conditions.

Programs

1. Observe a complete lightcurve over several nights. Obtain several lightcurves using differential photometry over several consecutive or near consecutive nights until complete rotational phase coverage is obtained. If no transformation to the standard magnitude system will be made, then no filter is necessary. This program will allow the period to be determined or serve as a check on a previous value.

This program will also show the amplitude of the lightcurve, which can differ from one opposition to the next owing to changes in the viewing geometry. If an asteroid is observed nearly pole-on (i.e. its rotational axis parallel to our line of sight) then very little or no variation would be seen. Conversely, if the rotational axis is perpendicular to our line of sight, we would be viewing at an equatorial aspect and would expect to see a maximum amplitude variation. Observations of a lightcurve amplitude over several oppositions can be used to estimate the direction of the asteroid's pole. Pole positions are known for very few asteroids but have an important consequence for asteroid collision models.

Observations from this type of program are very useful and are very likely to be published in the amateur journals and possibly in a professional journal, particularly if the results are new and the observations extensive.

2. Measurement of phase effects. The goal of this program requires that more lightcurves than in Program 1 be obtained which cover a range of nights before, during, and after the asteroid's opposition. This program requires all of the comparison stars used (and hence the asteroid) to be transformed to the standard magnitude system. A standard V filter is preferable for the observations although B is also acceptable.

Asteroids are known to become brighter as they approach their opposition even when all distance effects have been accounted for. For a phase angle (the Earth-asteroid-Sun angle) greater than or equal to seven degrees, the brightness of an asteroid (measured in magnitudes) increases linearly with phase angle as it approaches opposition. This increase is related to the asteroid's albedo. For phase angles less than seven degrees, the brightness increases more rapidly in a non-linear fashion. This extra brightening is known as the "opposition effect" and is believed to indicate the texture of the asteroid's surface.

Plots showing the distance corrected brightness of an asteroid versus phase angle are called phase curves. These have been made for relatively few asteroids. The slope of the linear part of the phase curve is called the phase coefficient. An observing program with lightcurves over a wide range of phase angles which result in a phase curve demonstrating the opposition effect and giving the measurement of the phase coefficient would be very valuable. The results would be worthy of publication in the amateur journals and in some cases, the professional journals.

3. Measurements of rotational color variations. The photometric colors of most bright asteroids have been well determined, however not all have been monitored for possible color variations through their rotational cycle. Very careful measurement of B-V and U-B color indices over several rotational cycles which consistently show color variations at the same rotational phase would be strong evidence for albedo spots on the asteroid's surface. Such variation has been confirmed only for 4 Vesta.

This program should be undertaken on asteroids with known rotational periods whose value should be checked as in Program 1. This program should only be attempted under very good conditions with instruments capable of 0.01 magnitude accuracy because the amplitude of any real variation will probably be only a few hundredths of a magnitude. Continuous UBV observations should be made of both the asteroid and the comparison star so that possible atmospheric or instrumental variations can be ruled out. The UBV observations should also be tied into the standard system.

Observations that consistently show significant color variations as well as observations showing no significant color variations greater than a few hundredths of a magnitude would be valuable contributions to the amateur and professional literature.

4. Determination of the pole position by astrometry. This program requires the most dedicated effort of all the others described. It is also much better suited for the amateur astronomer because he or she has one great advantage over the professional--the consistent availability of telescope time. This program involves obtaining lightcurves of a given asteroid over several months, starting roughly two months before opposition when the asteroid becomes available in the early morning sky. As many lightcurves on as many nights as possible should be obtained over the next four months or so as the asteroid passes through

opposition and then becomes lost in the evening sky. All lightcurves should be observed differentially with respect to a comparison star and later transformed to the standard system. This will allow the goals of Program 2 to be accomplished in addition to the ones described below.

Numerous lightcurves throughout one apparition allow a newly refined technique known as "photometric astrometry" to be applied to determine the sidereal period and pole position of the asteroid. These extensive observations would also allow changes in the amplitude and shape of the lightcurve over different phase and aspect angles to be studied. Such an observing program has only been attempted once by professional astronomers who observed the asteroid 44 Nysa. Details describing this technique may be found in Taylor and Tedesco (1983).

An extensive program such as this would be best undertaken by a group of amateurs with access to the same equipment. Any future professional campaigns on a particular asteroid are likely to be mentioned in the amateur literature and amateur observations would undoubtedly be welcomed. Because applying the technique of photometric astrometry in the final reduction of many lightcurves is complex, it would probably be wise to seek professional assistance.

III. SELECTION OF PROGRAM OBJECTS

Table 1-1 lists presently known lightcurve parameters for the 75 asteroids which can reach an opposition V magnitude brighter than 10.5. These data are taken from the Tucson Revised Index of Asteroid Data (TRIAD) lightcurve file maintained by E. F. Tedesco and published in the book Asteroids (see Section VIII). The table gives the asteroid's number and name, synodic rotational period (in hours), the range in amplitude which its lightcurve has been observed to exhibit (in magnitudes), and a quality code for the period determination. Quality 0 indicated that the asteroid has been previously observed but that the observations were not sufficient for a period determination to be made. Quality 1 indicates the given period is in doubt and that new observations are needed. A 2 indicates the given period is probably correct although observations to verify and improve the period would be worthwhile. Quality 3 indicates the result for the period is secure while a 4 indicates the given period is the sidereal period and that additional lightcurve observations are not needed for further study. The next column gives the number of oppositions during which the asteroid has been observed and the final column gives the phase coefficient (in magnitudes/degree), if known. Asteroids which do not have periods listed or known have been observed during only one opposition represent prime targets for observation.

An ephemeris is necessary to determine which of these asteroids are near opposition at any given time as well as to determine their apparent magnitudes. Many asteroids listed in the table can have opposition magnitudes fainter than V magnitude 10.5

TABLE 1-1

Parameters for Asteroids Reaching Opposition Magnitude Brighter than V 10.5

ASTEROID	PERIOD	AMPLITUDE	Q	NOP	COEFF
1 CERES	9.078	0.04	3	3	0.036
2 PALLAS	7.8106	0.12 - 0.15	4	9	0.037
3 JUNO	7.213	0.15	2	3	0.025
4 VESTA	5.34213	0.10 - 0.14	3	9	0.025
5 ASTRAEA	16.81184	0.21 - 0.27	4	4	0.01
6 HEBE	7.27445	0.06 - 0.19	4	4	0.027
7 IRIS	7.135	0.04 - 0.29	3	7	0.029
8 FLORA	13.6	0.01 - 0.04	2	2	0.028
9 METIS	5.064	0.06 - 0.26	2	4	0.04
10 HYGIEA	18.	0.09 - 0.21	2	3	0.03
11 PARTHENOPE	10.67	0.07 - 0.12	2	3	
12 VICTORIA	8.654	0.20 - 0.33	3	2	
13 EGERIA	7.045	0.12	2	1	
14 IRENE	9.35	0.04 - 0.10	2	3	
15 EUNOMIA	6.0806	0.42 - 0.53	3	6	0.038
16 PSYCHE	4.303	0.11 - 0.32	3	2	0.021
17 THETIS	12.275	0.12 - 0.36	2	2	
18 MELPOMENE	11.572	0.15 - 0.35	3	3	
19 FORTUNA	7.46	0.25	2	1	
20 MASSALIA	8.0980	0.17 - 0.24	3	3	0.031
21 LUTETIA	6.133	0.15	2	1	
22 KALLIOPE	4.147	0.14 - 0.30	2	3	0.031
23 THALIA	12.30	0.10 - 0.19	2	3	
25 PHOCAEA	10.	0.18	1	2	
27 EUTERPE	8.500	0.15	2	1	
28 BELLONA	15.695	0.03 - 0.19	2	2	
29 AMPHITRITE	5.390	0.11 - 0.13	2	6	0.032
30 URANIA	13.668	0.14	2	1	
31 EUPHROSYNE	5.54	0.09	2	1	
33 POLYHYMNIA	18.601	0.14	2	1	
37 FIDES	7.33	0.10 - 0.18	2	2	
39 LAETITIA	5.1382	0.18 - 0.54	3	7	0.026
40 HARMONIA	9.1358	0.22 - 0.28	2	2	
41 DAPHNE	5.9878	0.38	3	1	0.053
42 IRIS	13.59	0.32	2	2	
43 ARIADNE	5.7506	0.13 - 0.70	3	2	0.048
44 NYSA	6.421416	0.22 - 0.48	4	5	
46 HESTIA	20.5	0.11	2	1	
51 NEMAUSA	7.785	0.14	2	2	
52 EUROPA	11.2582	0.09	2	1	
54 ALEXANDRA	7.05	0.12	2	1	
63 AUSONIA	9.297	0.47	3	1	0.035
68 LETO	14.848	0.15 - 0.19	3	2	
78 DIANA	8.	0.15	1	1	
79 EURYNOME	5.979	0.05	3	1	0.032
80 SAPPHO	14.05	0.07 - 0.37	2	2	
88 THISBE	6.0422	0.19	2	1	
89 JULIA	11.3872	0.25	3	3	0.035
97 KLOTHO	35.	0.05 - 0.25	2	2	
115 THYRA	7.244	0.20	2	1	

TABLE 1-1 (continued)

ASTEROID	PERIOD	AMPLITUDE	Q	NOP	COEFF
129 ANTIGONE	4.9572	0.32	3	2	0.024
135 HERTHA	8.40	0.15	2	2	
144 VIBILIA					
182 ELSA	85.	0.6	2	1	
192 NAUSIKAA	13.622	0.42	3	1	0.040
194 PRONKNE	9.5	0.18	1	1	
216 KLEOPATRA	5.394	0.40 - 1.20	3	3	
230 ATHAMANTIS	7.996	0.10	2	1	
324 BAMBERGA	29.43	0.07	2	2	
344 DESIDERATA	13.	0.2	1	1	
349 DEMBOWSKA	4.7012	0.30 - 0.40	3	2	0.022
354 ELEONORA	4.2772	0.14 - 0.30	3	3	0.02
387 AGUITANIA	24.	0.09	1	2	
393 LAMPETIA	38.7	0.14	2	1	
419 AURELIA		0.10	0	1	
433 EROS	5.2703	0.00 - 1.50	4	3	0.024
471 PAPAGENA	7.113	0.12	2	1	
511 DAVIDA	5.17	0.06 - 0.25	2	4	
532 HERCULINA	9.406	0.08 - 0.18	2	3	
554 PERAGA	13.63	0.22	2	1	
654 ZELINDA	31.90	0.3	2	1	
704 INTERAMNIA	8.723	0.05 - 0.11	3	2	0.044
747 WINCHESTER	8.	> 0.1	1	1	
1036 GANYMED					
1627 IVAR					

if the opposition is unfavorable. Positions and magnitudes for all numbered asteroids which come to opposition during a calendar year are published annually by the Institute for Theoretical Astronomy, Leningrad, U.S.S.R. This book, Ephemerides of Minor Planets, may be found at most major observatories and in the astronomical collection of libraries at most universities.

Another source for ephemeris positions and even finding charts is a bimonthly newsletter, "Tonight's Asteroids." This publication is available free and one only need write Dr. J. U. Gunter, 1411 N. Mangum Street, Durham, N.C. 27701. Be sure to include a self-addressed stamped envelope and a brief letter as to your interests.

Ephemerides for asteroids of special interest or for which observations are especially encouraged are often given in the Minor Planet Bulletin, a publication described in Section VII. Ephemerides for specific asteroids that can not be obtained from other sources may be obtained by writing to The Minor Planet Center, Smithsonian Astrophical Observatory, 60 Garden Street, Cambridge, MA 02138, or by writing the author. Observers planning a serious, ambitious program for many asteroids who do not have an ephemeris available may request a copy by sending a letter of intent to

Dr. Brian Marsden at the Minor Planet Center. Very few copies of this book exist, ergo individuals obtaining a copy should make the best use of it.

The magnitudes of asteroids given in the above mentioned book are their B magnitudes. The V magnitude of an asteroid is roughly 0.7 magnitudes brighter. Determining whether or not a particular asteroid is bright enough to be observed with a given instrument depends upon the observing program to be undertaken. Previous experience or test measurements of stars of similar brightness will show whether or not measurements with the necessary accuracy can be made.

IV. OBSERVING PROCEDURE

The first step in finding an asteroid is to prepare or obtain a finding chart. The charts constructed by Dr. Gunter (described in Section III) are excellent and very convenient to use. If a chart for the desired asteroid is not available from there, one may be constructed using an ephemeris and a good star atlas. The star atlas should have limiting magnitude which is at least as faint as the asteroid. Good choices are the Smithsonian Astrophysical Observatory Atlas, the new AAVSO Variable Star Atlas, or Hans Vehrenberg's Photographic Atlas. It is best to construct the finding chart on a photocopy of the star chart simply by making small tick marks on the chart at the ephemeris positions. A smooth curve through the points shows the path of the asteroid along which positions between ephemeris dates may be interpolated. A sample finding chart which was constructed by Dr. Gunter using the Hans Vehrenberg Atlas is shown in Figure 1-1.

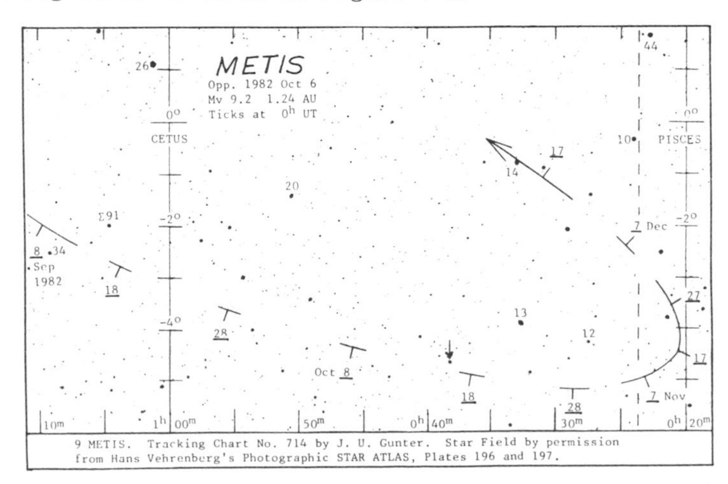

Figure 1-1. A typical finding chart.

To find the asteroid, locate the star field where the asteroid is suspected to be and begin matching up the stars in the field of view to those on the finding chart. An extra "star" near the predicted position is the asteroid. After several minutes you may be able to detect the asteroid's motion which will confirm your "discovery."

The next step is to locate a suitable comparison star. It is ideal to choose a star of nearly the same color and brightness as the asteroid which is less than one or two degrees away. This will allow differential extinction to be ignored in the reduction process for all observations made at an altitude greater than 35 degrees above the horizon. If possible, choose a star which is a member of the SAO Catalog so that it may be easily identified. The color match may be done in advance by choosing a comparison star with nearly solar colors (spectral type G2) from the SAO Catalog. Otherwise, the comparison star may be chosen at the telescope, comparing the instrumental colors (primarily B-V) of the asteroid to the candidate stars. Since a nearly perfect match is rarely obtained, choose the best match available which is still within two degrees of the asteroid. It is ideal if a star for comparison can be chosen which is ahead of the asteroid in the direction of its motion. This star may then be suitably close to the asteroid for use as a comparison for several nights. If a comparison star is to be used on several nights and especially if the observations will later be transformed to the standard system, it is best to make occasional measurements of another "check star" to monitor the constancy of the comparison.

Choose a filter appropriate for the observing program to be undertaken. Differential photometry for lightcurves which will not be transformed do not require a filter, while observations which will be transformed require either a V or B filter. In general, choose the filter (or no filter) which gives you the largest signal from the asteroid.

Since most asteroids are faint, a small aperture should be used in order to eliminate as much background skylight as possible. So choose the smallest aperture which is convenient to use but which is large enough to prevent loss of light beyond the edges which could render measurements useless. The aperture should also be large enough to allow objects to be centered easily so that measurements may be made fairly quickly.

If the photometer is a photon counting device, use integration times long enough to obtain at least 10,000 counts from the asteroid alone (after sky has been subtracted out). The same requirement is true for the comparison star. This will give measurements which can be accurate to 0.01 magnitude. For other photometer systems, strive to obtain a signal-to-noise ratio of at least 100 or to obtain readings accurate to one percent. These will translate to accuracies on the order of 0.01 magnitude.

The recommended observing sequence is:

$$S_c \; C \; S_c$$

$$S_a \; A \; S_a$$

$$S_c \; C \; S_c$$

$$S_a \; A \; S_a$$

°
°
°

C and A refer to measurements of the comparison star and asteroid while S_c and S_a refer to measurements of the sky in the immediate vicinity of the comparison star and asteroid, respectively. Since the asteroid may be faint, be sure that faint stars are excluded from the aperture when making sky measurements. If observations are being made in more than one filter, obtain a measurement in each desired filter for all S, C, and A observations. If observations are to be interrupted or terminated, they should be ended with the measurements of the comparison star and its sky. This will ensure that all asteroid measurements are bracketed by comparison star observations which were made at very nearly the same time. This is very important in determining whether observed variations in the asteroid are real or due to atmospheric or instrumental effects. When making the measurements, record each sky value and the values for C and A. Also record the Universal Time (UT) of each C and A measurement to the nearest minute. An example of one night's observations is given in Section IV.

The shorter the cycle time between the asteroid observations, the better. This will give higher resolution of the lightcurve and makes its analysis easier. However, do not sacrifice accuracy for speed, as accurate observations are the goal. Typical cycle times will be on the order of 10 to 20 minutes, depending on the instrumentation. Initial cycle times may be longer until a routine is established. Continue to cycle for as long as possible in order to obtain coverage of as much of the lightcurve as possible.

If standard stars are observed in order to make transformation to the standard system, it is best to also choose standard stars of nearly solar colors. This will minimize the effects of second order color terms in the reduction and may allow them to be ignored. As discussed in Section V, standard star observations for the transformation may be done on a separate night. A very useful and easily found star catalog is the Arizona-Tonanzintala Catalog published in _Sky and Telescope_ circ. 1965.

V. REDUCTION PROCEDURES

Differential Lightcurves

The first step in the reduction of any photometric data is the subtraction of the sky observations. Take the average of the two sky measurements made immediately before and after each C and A measurement and subtract it from the C or A measurement to obtain sky corrected measures denoted as C_o and A_o. If different integration times were used for the S, C, or A measurements, ensure that the units are all the same before performing the reduction, e.g. make sure all measurements are in counts/second or counts/10 seconds, etc.

Each asteroid measurement should have been immediately preceded and followed by measurements of the comparison star. Take the average of these two bracketing comparison star observations (which have been corrected for the sky) and call this value $\overline{C}_o$. The differential magnitude, ΔM, between the asteroid measurement and its bracketing comparison star measurements may be found from:

$$\Delta M = 2.5 \text{ Log}(\overline{C}_o/A_o)$$

where the sense of the difference is the magnitude of the asteroid minus the magnitude of the comparison star. An example of this reduction procedure is presented in Section VI.

A plot of ΔM versus time reveals the lightcurve of the asteroid. The value of the time used for the abscissa should be the time of the asteroid observation. Note that when constructing a lightcurve plot, the value of ΔM increases downward. Thus, if the asteroid increases in brightness, the lightcurve increases upward even though the value of ΔM decreases. (See Figures 1-3 and 1-4 in Section VI.)

The constant monitoring of the comparison star allows the small scale variations due to changes in sky conditions or instrumental sensitivity to be smoothed out of the asteroid lightcurve. However, an asteroid observation for which the comparison star measurements are vastly different should be treated with caution and probably deleted from the lightcurve. Such a rapid change due to some external factor may have also affected the asteroid measurement and an average between the two comparison star observations is unlikely to have smoothed out the variation. On the other hand, if variations are seen in the asteroid measurements while the comparison star remains constant, the variations are almost certainly real.

Reductions to the Standard Magnitude System

This procedure is beyond the scope of this chapter. Since this transformation requires observations of many standard stars under good photometric conditions, it may be heartening to learn that this need not be done on the same night as the lightcurve observations. This is because all of the asteroid observations are referenced to a comparison star. Once the comparison star's magnitude is known, the asteroid's magnitude on each night it was observed with respect to that comparison star may be determined.

Observing the standard stars for the transformation on the same night as the lightcurve observations has the advantage that the comparison star, itself, may be used as an extinction star to determine the extinction coefficients. If the transformation is not performed on the same night as the lightcurve observations, then the check star should also be measured when the transformation is being done and the differential magnitudes should be compared between the two nights. This will serve as a check on the non-variability of the comparison star. If asteroid observations are made over a long time interval involving many comparison stars, it is best to transform all of the comparison stars on the same night to assure uniformity between them.

Reduction of Phase Curves

After the comparison star and asteroid magnitudes have been transformed, the apparent magnitudes of the asteroid need to be corrected for relative Earth-asteroid and Sun-asteroid distances. This correction is made by computing the absolute magnitude of the asteroid, $V(1, \alpha)$, defined as the magnitude the asteroid would have if it were at a hypothetical position exactly 1 A.U. from the Earth and Sun but observed at the same phase angle, α. $V(1, \alpha)$ is calculated from:

$$V(1, \alpha) = V - 5 \, \mathrm{Log}(r \, \Delta)$$

where V is the apparent magnitude observed for the asteroid, r is the asteroid-Sun distance, and Δ is the Earth-asteroid distance. (Note: Even though the variable names refer to V observations, the reduction for B observations is identical.) The value for V must be chosen from the same feature in all the lightcurves. Using one of the distinct maxima which is recognizable as the same feature in all or most of the lightcurves is usually best. Values for r, Δ, and α for bright asteroids may be found in the *Ephemerides of Minor Planets*, where the variable name for the phase angle is called β. If this book is not available or does not contain these quantities for the reduction of a given asteroid, the values may be obtained from the author.

A plot of $V(1, \alpha)$ versus α should reveal a phase curve. Traditionally, the numerical value of $V(1, \alpha)$ increases downward and the phase angle increases to the right. Figure 1-2 displays points on a theoretical phase curve. The slope of a least squares fit to the linear portion of the phase curve (α greater than or equal to 7 degrees) gives a measurement of the phase coefficient. Values for this important parameter typically range between 0.02 and 0.05 magnitudes/degree. Data showing the non-linear brightening towards zero degrees phase angle due to the opposition effect are also very important to the study of asteroid surface properties.

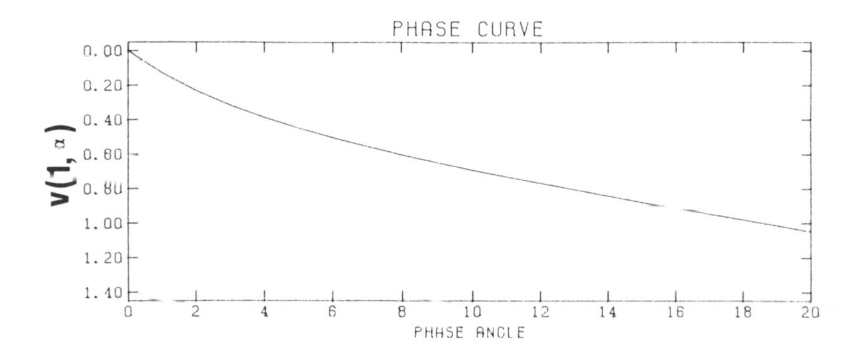

Figure 1-2. A theoretical phase curve.

Reduction of Color Variation Measurements

Differential magnitudes between the asteroid and the comparison star in each filter should first be computed. Thus, for each asteroid observation, there should be a value for ΔV, ΔB, and ΔU. A plot of ΔV versus time should be made to show the lightcurve. A set of color indices (which are not yet transformed on to the standard system) should then be determined from:

$$(B-V) = \Delta B - \Delta V$$

$$(U-B) = \Delta U - \Delta B$$

A plot of (B-V) and (U-B) versus time should be made using the same scale as the lightcurve and examined for significant variations. Since these values were determined in a doubly differential technique they should be insensitive to small changes in transparency or instrumental sensitivity. Regardless, this type of observation should only be carried out on good photometric nights. Any significant color variations should be referenced to particular features in the lightcurve. Before these variations can be considered definitive, observations from one or more additional nights will need to be made to show that the color variations occur at the same relative place in the lightcurve.

VI. AN EXAMPLE

The data used in this example are real measurements made by the author and P. Hartigan using a DC photometer with an uncooled RCA 1P21 photomultiplier and a standard V filter. The telescope was a 32 cm Cassegrain constructed by S. W. Schultz at the Macalester College Observatory, St. Paul, Minnesota. All measurements are read directly off a microammeter. The asteroid, 44 Nysa, was near V magnitude 10.0 at the time of the observations.

Table 1-2 presents the data and the reduction steps for the observations on the night of September 25, 1979. All readings are expressed in units of 10^{-11} amperes. Observations were made using the observing sequence given in Section IV, and the reductions were performed as outlined in Section V. The first five columns give the observed values in the order they were taken, where the listed time is when the comparison star (C) or asteroid (A) was measured. The reductions are shown in the next four columns where the mean for each of the two sky measurements was first calculated and then subtracted from the corresponding C and A. The mean of the two sky corrected comparison star measurements which bracketed each asteroid observation, $\overline{C}_o$, was then calculated. The final column gives the value of ΔM which was calculated using this comparison star mean and the sky corrected asteroid measurement using the equation given in Section V.

A plot of ΔM versus time is given in Figure 1-3 which reveals a lightcurve for 44 Nysa having a 0.4 magnitude amplitude. Two minima and one complete maximum are displayed as well as parts of a second (and possibly a third) maximum. Because nearly all asteroid lightcurves display two maxima and minima per rotation, we conclude that this lightcurve over six hours of time nearly covered one complete rotation. We can not be sure that the width of the second maximum, which is partially observed at the beginning and end of the night, has been completely observed. Thus, we estimate that the rotational period of 44 Nysa is slightly longer than six hours.

Figure 1-4 shows a lightcurve obtained on the following night with the same equipment and comparison star. By plotting both lightcurves on tracing paper and superimposing, we see that the

TABLE 1-2

Example Data and Reduction for Asteroid 44 Nysa

OBJ.	UT	FIRST SKY	OBJECT READING	SECOND SKY	MEAN SKY	SKY CORR OBJECT	MEAN COMP	DELTA MAG.
C	3:05	51	213	49	50.0	163.0		
A	3:10	50	134	50	50.0	84.0	165.0	.73
C	3:15	50	217	50	50.0	167.0		
A	3:20	49	138	47	48.0	90.0	166.5	.67
C	3:25	48	214	48	48.0	166.0		
A	3:30	48	135	48	48.0	87.0	174.0	.75
C	3:40	48	230	48	48.0	182.0		
A	3:55	47	122	47	47.0	75.0	184.5	.98
C	4:00	48	234	46	47.0	187.0		
C	4:40	47	226	47	47.0	179.0		
A	4:45	48	126	48	48.0	78.0	178.5	.90
C	4:50	48	227	50	49.0	178.0		
A	4:55	49	124	49	49.0	75.0	177.0	.93
C	5:00	48	224	48	48.0	176.0		
A	5:05	48	130	48	48.0	82.0	177.5	.84
C	5:08	48	226	46	47.0	179.0		
A	5:10	46	130	46	46.0	84.0	178.5	.82
C	5:12	46	224	46	46.0	178.0		
A	5:15	46	134	44	45.0	89.0	177.0	.75
C	5:20	45	221	45	45.0	176.0		
A	5:25	46	136	44	45.0	91.0	171.0	.68
C	5:30	45	211	45	45.0	166.0		
C	6:00	44	211	44	44.0	167.0		
A	6:05	44	131	43	43.5	87.5	166.5	.70
C	6:10	43	209	43	43.0	166.0		
A	6:20	43	128	41	42.0	86.0	164.5	.70
C	6:25	42	205	42	42.0	163.0		
A	6:35	42	122	42	42.0	80.0	166.0	.79
C	6:40	40	210	42	41.0	160.0		
A	6:45	41	118	41	41.0	77.0	165.0	.83
C	6:50	40	201	40	40.0	161.0		
A	7:00	41	115	40	40.5	74.5	163.5	.85
C	7:10	40	206	40	40.0	166.0		
A	7:15	40	103	40	40.0	63.0	157.5	.99
C	7:20	39	188	39	39.0	149.0		
A	7:25	40	102	38	39.0	63.0	157.5	.99
C	7:28	39	205	39	39.0	166.0		
A	7:30	38	96	38	38.0	58.0	160.5	1.00
C	7:35	38	193	38	38.0	155.0		
A	7:40	39	102	39	39.0	63.0	162.0	1.03
C	7:45	39	207	37	38.0	169.0		
A	7:50	38	101	38	38.0	63.0	162.0	1.03
C	7:55	38	193	38	38.0	155.0		
A	8:00	36	105	38	37.0	68.0	160.5	.93
C	8:05	37	203	37	37.0	166.0		
A	8:10	35	105	35	35.0	70.0	154.5	.86
C	8:20	35	177	33	34.0	143.0		
A	8:25	33	107	33	33.0	74.0	148.5	.76
C	8:30	34	188	34	34.0	154.0		
A	8:50	33	117	33	33.0	84.0	158.5	.69
C	8:55	33	196	33	33.0	163.0		
A	9:00	32	117	34	33.0	84.0	157.0	.68
C	9:05	34	184	32	33.0	151.0		

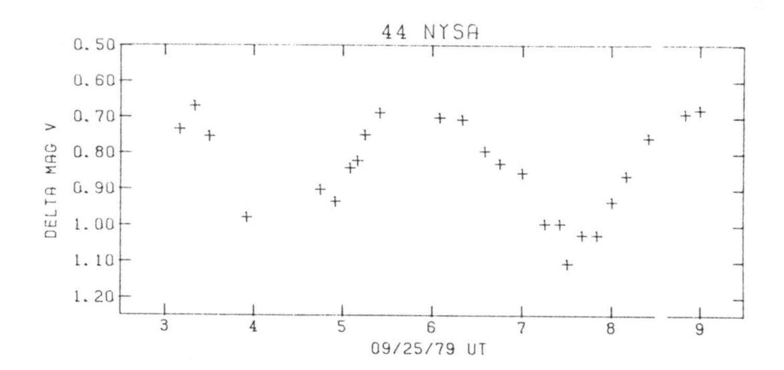

Figure 1-3. Photoelectric lightcurve of Asteroid 44 Nysa, September 25, 1979.

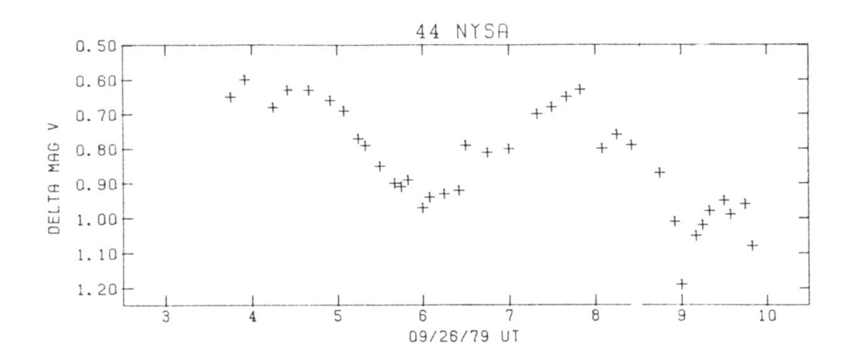

Figure 1-4. Photoelectric lightcurve of Asteroid 44 Nysa, September 26, 1979.

best match shows that the maximum at 6:00 on 25 September corresponds to the one observed at 7:40 on 26 September. (The match of the vertical scale is arbitrary because we have not accounted for phase and distance effects.) This implies that an integer number of rotational cycles had elapsed in the 25.667 hours separating these maxima. Our estimated period of slightly more than six hours allows us to conclude that 4 cycles elapsed in this time. Dividing 25.667 by 4 gives us a more accurate determination for the rotational period equal to 6.42 hours.

To determine our error we slide the tracing paper until the worst tolerable superposition is obtained. For these lightcurves this corresponds to an elapsed time of 25.83 hours. Dividing this by 4 gives a period equal to 6.45 hours. Thus we conclude from our data that the period is 6.42 ± 0.03 hours. Our probable error can be decreased by observing the lightcurve again many nights later as uncertainties in the superposition of the curves will become smaller with respect to the elapsed time.

VII. PUBLICATION OF OBSERVATIONS

Most amateur observations of asteroids are published by the American Lunar and Planetary Observers (ALPO) Minor Planets Section in their quarterly publication, the Minor Planet Bulletin. Dr. Fredrich Pilcher, Illinois College, Jacksonville, IL 62650, is the head of the Minor Planets Section and edits the Bulletin. Even though this publication is primarily for amateurs, it is widely read by and contains contributions from professional astronomers. In some cases, collaborating observations with professional astronomers lead to results published in one of the professional journals. The primary professional journal which contains papers on asteroid studies is Icarus, a monthly journal published by the Academic Press, New York.

VIII. FURTHER READING

Persons interested in observational studies of asteroids are encouraged to join the ALPO Minor Planets Section by writing Dr. Fredrich Pilcher at the address given in Section VII. Membership in the Minor Planets Section includes a subscription to the Minor Planet Bulletin which contains information on current observing programs.

Must reading for persons interested in all aspects of the study of asteroids is the book, Asteroids, edited by Tom Gehrels. This book was the result of an international conference on asteroids held in Tucson in 1979. It contains review articles on all aspects of asteroid study allowing even a new reader in the field to learn about the current state of asteroid research. It also contains a listing of all the known parameters for asteroids which is called the Tucson Revised Index of Asteroid Data (TRIAD). More recent articles which update some of the data in the TRIAD file are primarily found in the journal Icarus. A number of the references below are examples of published asteroid observations by amateurs.

REFERENCES

Arizona-Tonanzintala Catalogue (1965). Sky and Telescope 30, 25.

Binzel, R. P., Harris, A. W. (1980). "Photoelectric Lightcurves of Asteroid 18 Melpomene." Icarus 42, 43.

Birch, P. V., Tedesco, E. F., Taylor, R. C., Binzel, R. P., Blanco, C., Catalano, S., Hartigan, P., Scaltriti, F., Tholen, D., Zappala, V. (1983). "Lightcurves and Phase Function of Asteroid 44 Nysa During Its 1979 Apparition." Icarus, 54, 1.

Cunningham, C., Kaitting, M. (1982). "Photoelectric Photometry of Asteroid 18 Melpomene." Minor Planet Bull. 9, 1.

Cunningham, C. (1983). "Photoelectric Photometry of 532 Herculina." Minor Planet Bulletin, 10, 4.

Giffert, M., Hoffman, M. (1981). "Observations of Asteroid 4 Vesta." Minor Planet Bulletin 8, 17.

Grossman, M., Hoffman, M., Duerbeck, H. (1981). "Photoelectric Measurements of 216 Kleopatra." Minor Planet Bulletin 8, 14.

McFaul, T. G. (1981). "Photoelectric Photometry of the Asteroid 980 Anacostia." Minor Planet Bulletin 8, 19.

Mims, S., Wallentinsen, D., and James, R. (1981). "Photoelectric Photometry of 68 Leto." Minor Planet Bulletin 8, 27.

Porter, A., Wallentinsen, D. (1980). "Minor Planet Rotations 1978-1979." Minor Planet Bulletin 7, 21.

Taylor, R. C., Tedesco, E. F. (1983). "Pole Orientation of Asteroid 44 Nysa via Photometric Astrometry." Icarus, 54, 13.

Welch, D., Binzel, R. P., Patterson, J. (1974). "The Rotation Period of 18 Melpomene." Minor Planet Bulletin 2, 20.

2. PHOTOMETRY OF PLANETS AND SATELLITES

G. Wesley Lockwood

I INTRODUCTION

The photometry of planets and satellites is, surprisingly, mostly quite recent, with the bulk of the important studies having been made during the 1970's. Following a brief survey in the UBV system by Harris (1961), photometry lay dormant until the advent of planetary exploration by spacecraft two decades later. Only then were the fundamental photometric properties of solar system objects determined. Many of these studies were carried out, at least for the brighter objects, with telescopes in the 24-inch class.

Comprehensive photometric surveys of the Galilean satellites of Jupiter were made in the UBV system by Millis and Thompson (1975) and in the *uvby* system by Morrison *et al*. (1974). A similar survey of Saturn's satellites was carried out by Noland *et al*. (1974). The surveys were aimed at answering three questions: (1) How bright are the satellites at opposition? (2) How do the satellites vary in brightness as they revolve about Jupiter?, and (3) What are the variations of brightness as a function of solar phase angle, i.e., the angle between Sun, Jupiter, and the Earth? The first question concerns the satellite albedos, that is, how much sunlight is reflected from their surfaces. The second describes the variation of albedo over the satellite's surface and is related to the actual composition (ice, rock, etc.) of the surface. Finally, the change in brightness as a function of solar phase angle tells us still more about the surface composition.

The geometry of the solar system constrains the range of solar phase angles seen from the Earth to 12° for Jupiter, 6° for Saturn, 3° for Uranus, and 2° for Neptune. Nonetheless, useful information can be obtained for each of these objects and their satellites through careful observations over the limited range of phase angle available.

While the Galilean satellites can always be observed visually (except during eclipse or while inside Jupiter's shadow), photoelectric photometry is possible only when they are located fairly near eastern or western elongation. Even so, proper subtraction of the sky light scattered from Jupiter's disk requires careful measurement.

Figure 2-1 shows how the brightnesses of the Galilean satellites vary as a function of solar phase angle, and Figure 2-2 shows the variations as a function of orbital phase angle. The orbital phase angle is measured from zero degrees at superior geocentric conjunction along the satellite orbit around the planet.

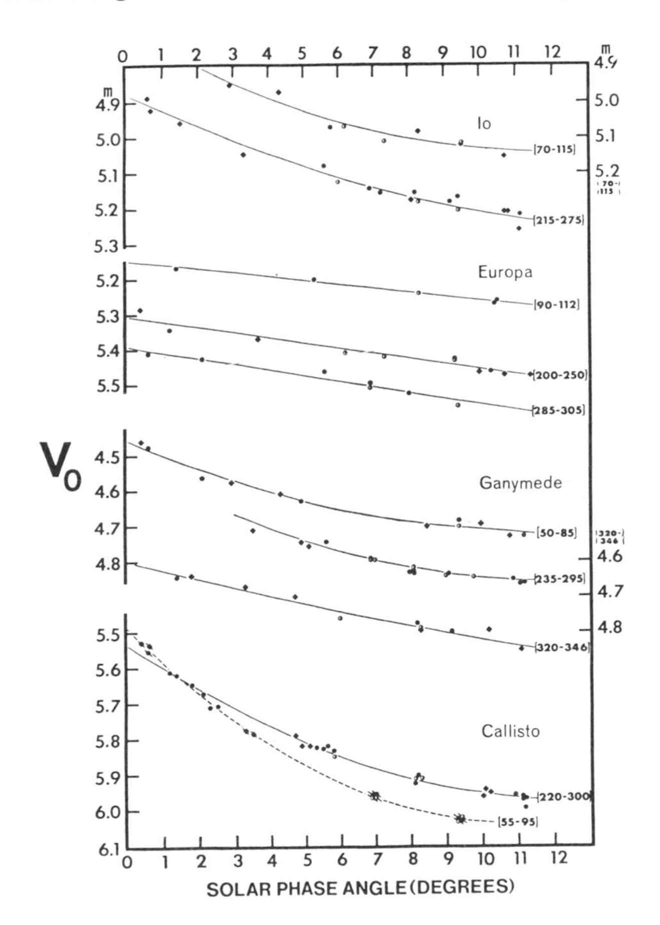

Figure 2-1. Variation of the V magnitude of the Galilean satellites as a function of solar phase angle for various faces of the Galilean satellites. Different symbols refer to data from two observatories in two different seasons. The small numbers in square brackets give the range of orbital phase angle over which the data for a particular curve was obtained. The equations for the curves for the bright faces are given in Table 2-3. This figure and Figure 2-2 (from Millis and Thompson 1975) are reprinted from Icarus, by permission.

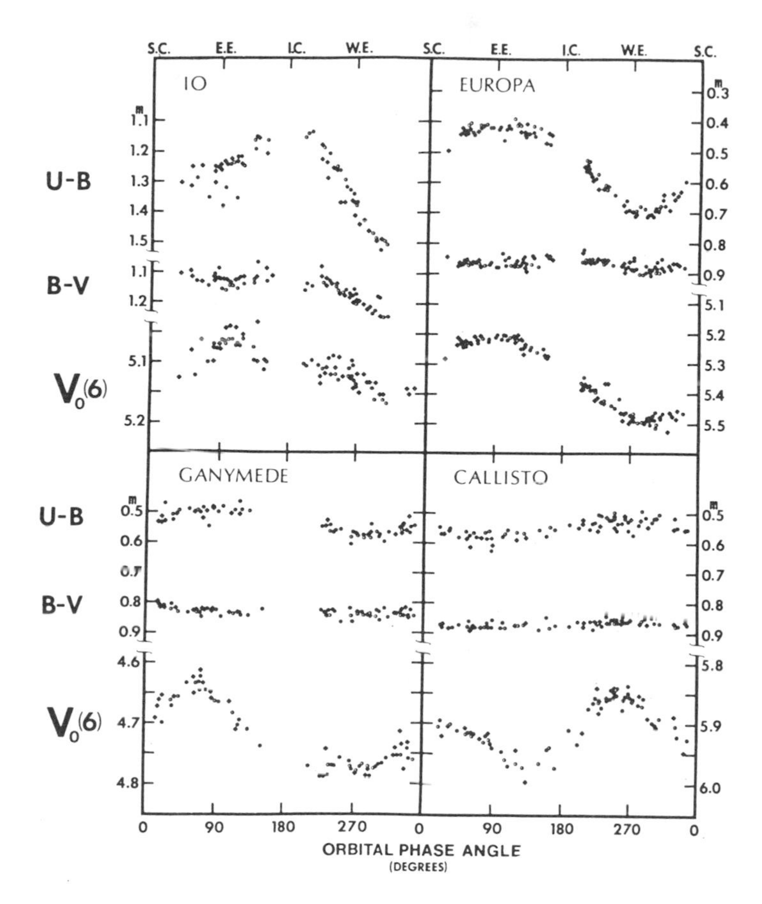

Figure 2-2. UBV magnitudes and colors of the Galilean satellites, corrected to mean opposition distance and a solar phase angle of 6°. Note the absence of data at conjunctions when the satellites are not visible from the Earth.

Another aspect of satellite and planetary photometry which has been studied concerns intrinsic variability. Of course, the brightnesses of all of these objects vary as the distances between Earth, Sun, and planet change. Thus we must normalize the observed magnitudes to some fixed distance, such as the mean opposition distance, before we can consider the question of intrinsic variability.

The variations of intrinsic brightness of several solar system objects have been measured since the early 1950's at the Lowell Observatory using a 21-inch telescope, in a program aimed at detecting small variations (less than one percent) of the brightness of the Sun. Variations of Titan (a satellite of Saturn) and Neptune have been shown to be possibly related to the solar cycle (Lockwood 1977; Lockwood and Thompson 1979). Figure 2-3 shows the long-term intrinsic variations of Titan, Uranus, and Neptune.

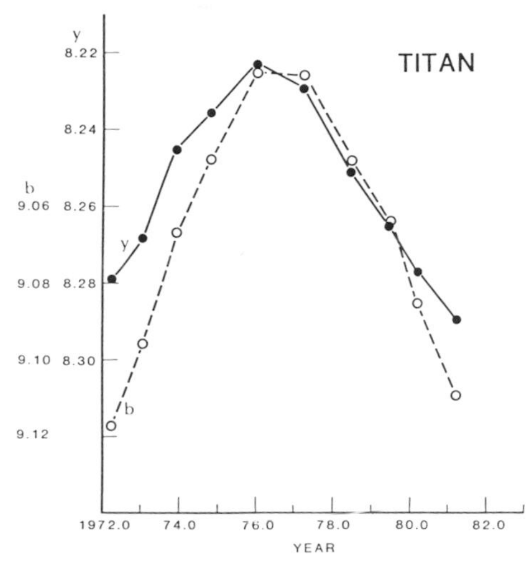

Figure 2-3a. Brightness changes of Titan in b (4700 Å) and y (5500 Å) measured at the Lowell Observatory.

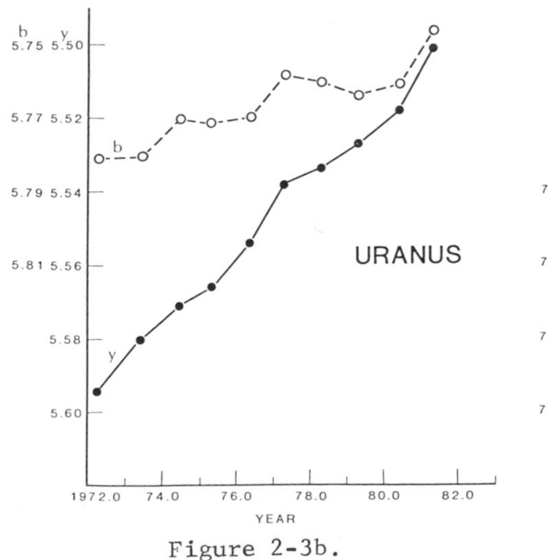

Figure 2-3b.
Brightness changes of Uranus.

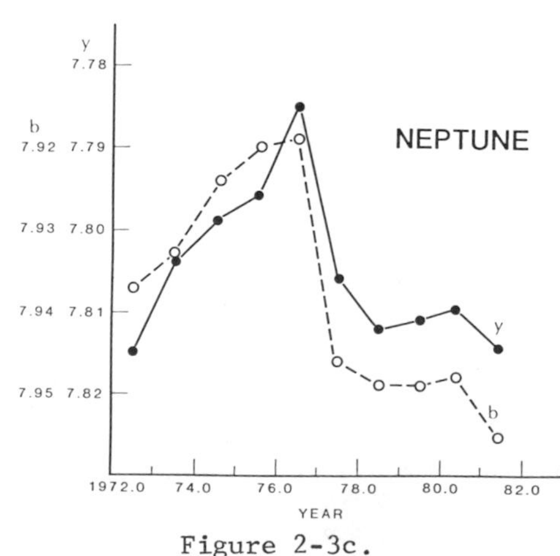

Figure 2-3c.
Brightness changes of Neptune.

These studies of the variability of planets and satellites were carried out much like the differential photometry of variable stars, where the program object is compared to one or two comparison stars of similar brightness nearby in the sky. However, since the planets continually move along the ecliptic, new comparison stars must be used every year or so. The approximate annual rates of motion are 30° for Jupiter, 12° for Saturn, 4° for Uranus, and 2° for Neptune. Obviously, the same comparison stars could be used for several years for Uranus and Neptune, since they move so slowly.

II. CHOOSING AN OBSERVING PROGRAM

A wide variety of possible programs can be carried out, encompassing a range of technical and observational complexity. At one extreme, an observer might wish simply to verify for himself known relationships such as the variation of brightness of planets and satellites with distance from the Earth and Sun, with solar phase angle, and with orbital phase angle. At the opposite extreme, an observer might be challenged to conduct a difficult original research program, for example, to improve the orbital and solar phase curves of the Galilean satellites, or to look for time variations of the brightnesses of Uranus or Neptune.

The programs suggested here concern objects whose telescopic images are sufficiently small that they can be measured just like stars. A table of solar system objects brighter than 10th magnitude is given below. Only Uranus and Neptune are perceptibly nonstellar in appearance, with angular diameters of 4 arcseconds and 2 arcseconds, respectively.

TABLE 2-1

Opposition V Magnitudes

Object	V mag
GALILEAN SATELLITES	
Io	4.9
Europa	5.3
Ganymede	4.6
Callisto	5.6
SATURNIAN SATELLITES	
Titan	8.3
Rhea	9.9
PLANETS	
Uranus	5.5
Neptune	7.9

Physical Studies of Planetary and Satellite Surfaces and Atmospheres

Solar Phase Curves. The solar phase functions shown in Figure 2-1 describe how the brightness of an object varies as the angle changes between the Earth, the object, and the Sun. For satellites, the solar phase angle is almost exactly the same as for the parent planet. The function depends upon the optical properties of the visible surface or atmosphere of the object. Objects with dense atmospheres such as Titan have a very weak dependence of brightness upon solar phase angle. However, those with solid ice or mineral surfaces, such as the Moon, Mars, the Galilean satellites, and the asteroids, have steep phase functions whose slopes change rapidly near opposition. For the Galilean satellites, a linear equation (straight line) is sometimes used for solar phase angles greater than 6°, but the whole curve from 0 to 12° is better described by a quadratic equation (parabola). Solar phase functions for the objects brighter than 10th magnitude are given in Table 2-3 in Section IV.

The verification of a solar phase function by making new observations is a feasible project for small telescopes equipped with a photoelectric photometer. Since the total range of variation for the Galilean satellites is several tenths of a magnitude, the phase function is fairly easy to measure. However, this is a difficult project for the objects Titan, Uranus, and Neptune, whose total range of variation due to the change in solar phase angle is less than 0.03 mag.

Orbital Phase Curves. Satellites such as the Galilean satellites of Jupiter also change in apparent brightness as they revolve about Jupiter (Figure 2-2). They are in synchronous rotation, which means that the same face always points to Jupiter, i.e., the satellite rotates with a period equal to the period of revolution. Consequently, areas of differing reflectivity come into view, which leads to the orbital lightcurve, or orbital phase curve, as it is often called. The peak-to-peak amplitudes of the orbital phase curves of the Galilean satellites are in the range 0.1 to 0.3 mag. Hence, these curves can be measured with fair accuracy by a careful observer.

Note that observations of the Galilean satellites taken at random times during an observing season will include a range of both orbital and solar phase angles. However, it is possible to schedule observations so that one effect dominates the observations. For example, if a satellite is always observed near elongation over a whole observing season, the solar phase angle will change from week to week, but the orbital phase angle will be relatively constant. On the other hand, if observations are carried out over a few nights close together, the orbital phase angle will differ from night to night, but the solar phase angle will not change very much.

The solar phase angles of the planets and the times of elongation or superior conjunction of the satellites are tabulated in the Astronomical Almanac (Superintendent of Documents, U. S. Government Printing Office, Washington, D.C. 20402, about $16 per year).

Secular Variability

Titan, Uranus, and Neptune have all shown secular, long-term variability, as shown in Figure 2-3. An extremely patient observer might consider a program to measure the variability over a period of several years.

Titan. The total observed range in brightness since 1972 is about 0.1 mag. The changes in brightness may be due to a cause-and-effect relationship between solar activity and the optical properties of Titan's atmosphere, or they may be related to the seasonal change in the aspect of Titan as viewed from the Earth. Since a full solar cycle is 22 years long and the period of Saturn's revolution is 28 years, only continued long-term observations will allow the two effects to be separated. An observer may expect to see variations no larger than 0.01 to 0.02 mag from one year to the next. Consequently, Titan should be regarded as a difficult object in terms of the accuracy required to see any variations.

Uranus. The long-term variability of Uranus is related to its changing aspect as seen from the Earth. Uranus is brighter seen pole-on (as in 1986) than when seen equator-on (as in 1966). The amplitude of variation is about 0.1 mag. There are two reasons for the variability. First, Uranus is oblate, with the equatorial diameter being about 3 percent larger than the polar diameter. When Uranus is seen pole-on from the Earth, its apparent diameter is larger and the planet appears brighter. Second, the albedo of the polar regions is slightly greater than that of the equatorial regions. Consequently, the planet will be brighter when seen pole-on, even if the oblateness effect is neglected.

Neptune. Small fluctuation in the brightness may be related to the solar cycle. The total amplitude observed since 1972 is about 0.04 mag. The brightness of Neptune was greatest around the time of solar minimum in 1976.

Io. Changes have also been observed in Io, due undoubtedly to the effects of its continuing volcanic activity. A total change of about 0.04 mag was observed between 1977 and 1978.

Europa, Callisto, and Rhea. The apparent long-term intrinsic changes in these objects seems to be very small, probably less than 0.005 mag.

Choosing Comparison Stars

In all of the programs described above, the observations have been made differentially, just as in variable star work. If possible, we prefer the comparison stars to be no more than 2 or 3 degrees away from the planet. This simplifies the task of setting the telescope and also means that the airmasses of the objects are all about the same.

There are two principal criteria for the selection of comparison stars, apart from proximity to the planet. First, we prefer the magnitudes of the comparison stars to be nearly the same as that of the planet or satellite. In practice, the magnitude differences are often less than 1 magnitude and nearly always less than 2 magnitudes. The effects of any possible nonlinearities in the data-recording electronics are thus minimized. Second, we want the color of the star to be similar to that of the Sun, to minimize undesirable color-term effects in the data reductions. In practice, we restrict comparison stars to spectral types between F5 (hotter than the Sun) to K0 (cooler than the Sun). Many field stars cooler than K0 turn out to be small-amplitude variables and are, hence, unusable. We use two comparison stars for each object. In this conservative approach, the stability of the photometer is assessed by looking at the differential magnitudes of the comparison stars, which should be constant from one night to the next and during each night. The planet or satellite can be compared independently to each comparison star alone. This approach differs somewhat from the common practice of observing a "check star" occasionally during the night.

For the Galilean satellites and Uranus, which are about 5th magnitude, suitable comparison stars can usually be found in the *Yale Catalogue of Bright Stars*. Fainter objects such as Titan and Neptune, which are about 8th magnitude require comparison stars tabulated in the *Smithsonian Astrophysical Observatory Catalog*, *Sky Catalog 2000.0*, or *Henry Draper Catalog*.

III. EQUIPMENT

The requirements for the telescope, photometer, and data acquisition equipment are quite modest. Long-term stability and reliability are essential, while state-of-the-art technical sophistication is somewhat secondary.

The Telescope

An aperture from 12 to 24 inches is suitable for programs involving the brighter solar system objects. Good tracking capability is essential since the objects must be kept centered in photometer aperture diaphragms on the order of 20 to 40 arcseconds in diameter.

Simple setting circles and a small finder are adequate for object acquisition.

The 21-inch telescope at Lowell, which has been used for solar system photometry for 30 years, is shown in Figure 2-4. Its size, construction, and instrumentation are in many ways similar to that of equipment used by advanced amateurs. The finder is a surplus elbow telescope with an aperture of about two inches. The setting circles are divided into 5-minute divisions for RA and 1° divisions for declination, with a vernier scale on each. The declination motion fine adjustment is made by means of a tangent screw and hand crank, and RA set/guide motions are controlled by speeding up or slowing down the stepper motor drive.

The Photometer

Our photometer is also shown in Figure 2-4. As is typical of modern, manually operated photometers, there is a wide-field viewer above the star diaphragm and a high-power guide eyepiece below the diaphragm, both brought into the telescope beam by means of movable mirrors.

The mechanical and optical requirements for the photometer are the same as for variable star work. There are six aperture diaphragms, ranging in size from 12 to 60 arcseconds, and a six-position filter wheel. For the Galilean satellites and Titan, we usually use a 20 arcsecond aperture, while Rhea requires a 12 arcsecond aperture. For Uranus large apertures are satisfactory. For Neptune we use a 40 arcsecond aperture in order to be able to include its 13th-magnitude satellite, Triton.

Filters

Standard colored glass UBV filters can be used for photometric programs involving planets and satellites. However, at Lowell we use intermediate-band <u>b</u> (4700 Å) and <u>y</u> (5500 Å) interference filters. These typically cost several hundred dollars each when manufactured to strict specifications; but sometimes bargains can be found for under $25 in the surplus stock of various manufacturers.

Electronics and Data Recording

The principal requirements for the electronics are long-term stability and linearity. Many amateur astronomers have solved these problems successfully with equipment of their own manufacture. We use a Pacific Photometric Model AD-6 amplifier-discriminator (Pacific Precision Instruments, 1040 Shary Court, Concord, California 94518), which comes with its own low- and high-voltage

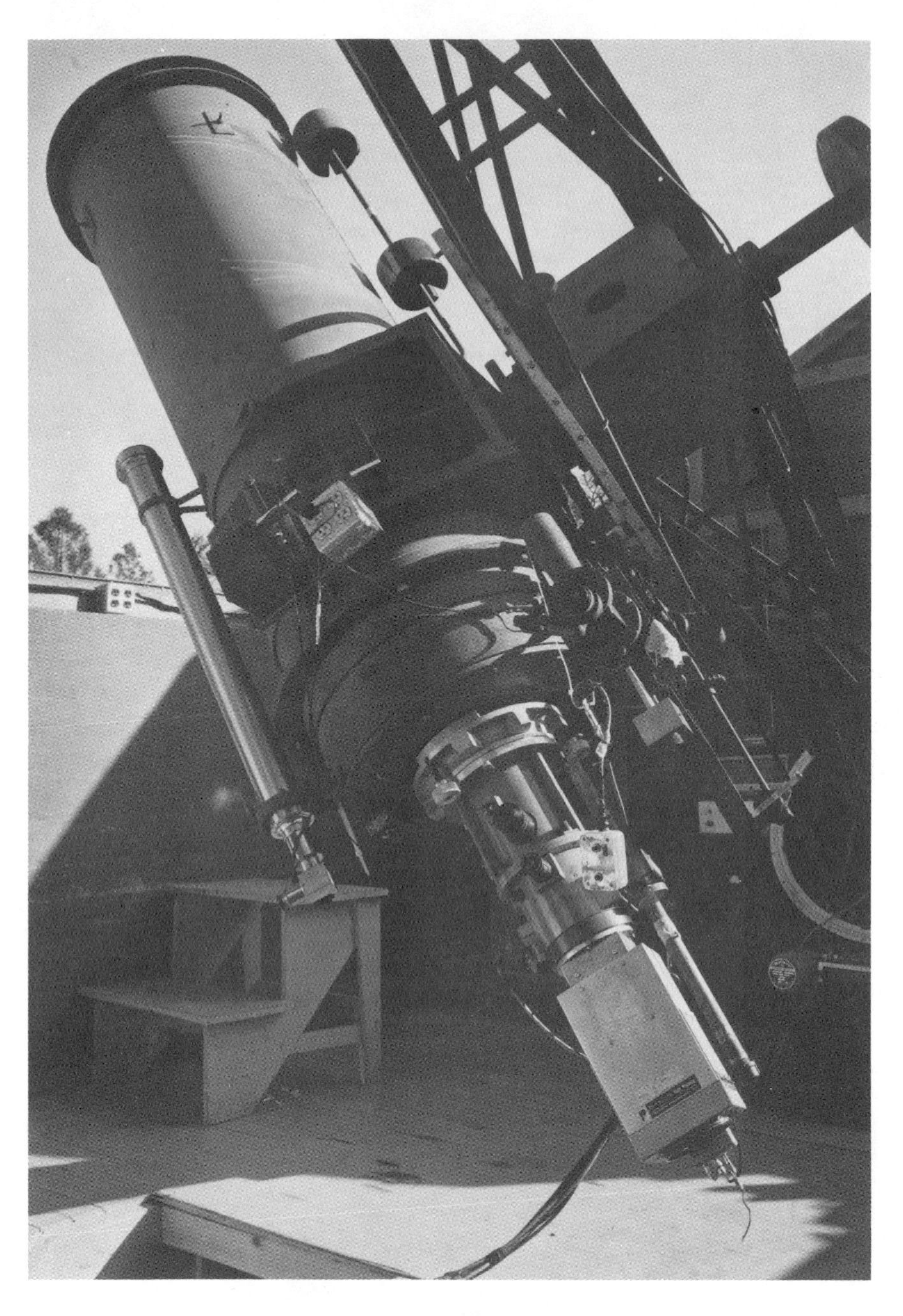

Figure 2-4. The 21-inch telescope and photometer at Lowell Observatory.

supply. The photomultiplier (an EMI 6256S) is housed in a thermoelectrically cooled chamber made by Products for Research (78 Holten Street, Danvers, Massachusetts 01923). The total cost of these commercially-made items is, unfortunately, fairly high, about $3,000.

We find it useful to record the following for each 10-second integration: local time, filter, star identification, star or sky, photon count. In our data system, all of these quantities are encoded and recorded automatically by an on-line PDP-11/15 computer with 4K memory which controls a Teletype ASR 33 teleprinter and paper tape punch. This equipment dates from about 1971 and will be replaced soon with a new computer and floppy disk system. Since many amateur astronomers already own computer equipment which is more modern and technically advanced than our present system, we need not make recommendations, but note that the cost of our present system (computer+Teletype, but exclusive of its counter and interface) is now only a few hundred dollars on the surplus market!

IV. OBSERVING TECHNIQUE

Sequence of Observations

We use an observing sequence which was devised by M. Jerzykiewicz of the Wroclaw University Observatory (Poland) for his variable-star work at the Lowell Observatory in 1972-1973. The scheme may appear arbitrary, but careful examination reveals a number of subtleties which experienced photometrists will appreciate:

1. Only one filter is used at a time for sequential measurements of the planet or satellite and the two comparison stars. Observations with each filter are therefore close together in time, which minimizes the effect of changing hour angle, airmass, instrumental sensitivity, sky transparency, etc.

2. Each object (planet and each comparison star) is measured an equal number of times. Hence, the statistical weights of various differential magnitudes are equal, providing that the planet and comparison stars are of nearly the same brightness.

3. Each observation consists of two sets of measurements separated by a sky measurement.

4. A self-contained "cycle" comprising measurements of two comparison stars and the planet through a single filter can be carried out in less than ten minutes. If sky conditions suddenly deteriorate, earlier cycles are not compromised. All cycles have equal statistical weight.

5. Observations are carried out (usually) within one hour of transit. Hour angle and flexure effects are minimal, and the airmass is a minimum.

A "cycle" consists of the following set of 10-second measurements all taken through the same filter, where "1" denotes the planet, "2" and "3" denote the comparison stars.

sky 2
sky 2
comp star 2
comp star 2
comp star 2

planet 1
planet 1
planet 1
sky 1
sky 1
planet 1
planet 1
planet 1

comp star 3
comp star 3
comp star 3
sky 3
sky 3
comp star 3
comp star 3
comp star 3

comp star 2
comp star 2
comp star 2
sky 2*
sky 2*

*Omitted if the subsequent cycle is observed through the same filter, since the cycle begins with "sky."

Successive cycles are arranged symmetrically. For example, if we wish to observe in two colors, b and y, with four cycles in each color, they will be arranged as:

cycle in y
cycle in y
cycle in b
cycle in b
cycle in b
cycle in b
cycle in y
cycle in y

Where each cycle is reduced separately to a set of differential magnitudes of the form "1"-"2", "1"-"3", "2"-"3".

How to Measure the Sky Brightness

The accuracy of photometric measurements of the Galilean satellites Titan and Rhea is limited by how accurately the sky can be measured. The reason for this is the high nonuniform background of scattered light from the planet within and near the field of the satellite.

It is convenient to imagine that the distribution of intensity of the scattered light from the planet is symmetrical about the planet and decreases radially outwards. We therefore estimate the intensity of the sky background at the position of the satellite observing it next to the satellite at the same radial distance from the planet, as shown in Figure 2-5.

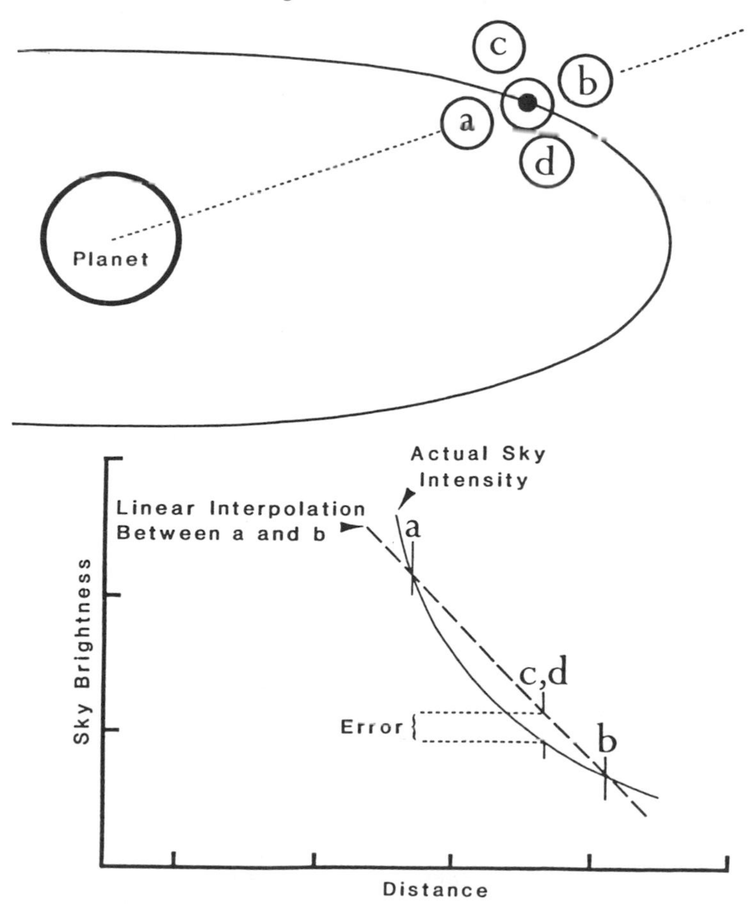

Figure 2-5. Sky brightness should be measured at locations c and d, whose distances from the planetary disk are equal.

Some observers will argue that "sky" should be measured at positions "a" and "s" in Figure 2-5, rather than "c" and "d," as indicated. However, this assumes erroneously that the falloff in sky brightness is linear with radial distance and that the average of measurements at "a" and "b" is therefore appropriate.

In satellite photometry, we always measure the sky at "c" and "d" during each cycle. The internal agreement of these measurements is one indication of how accurately the measurement has been made. In the case of Titan, using a 20 arcsecond aperture, the sky brightness is typically about 3 percent of the brightness of Titan. Hence, a 10 percent error in the sky measurement itself contributes only 0.3 percent to the uncertainty in the brightness of Titan.

One can easily visualize the following much more difficult observational condition. The sky brightness near Rhea at its elongation, observed with a 12 arcsecond aperture, is about 10 percent of the brightness of Rhea itself. The uncertainty in measuring the sky owing to the proximity of Rhea to Saturn's rings and disk is approximately 20 percent at best. Hence, the uncertainty due to sky cancellation thus approaches the unacceptable level of 2 percent or 0.02 mag.

V. REDUCTIONS

We will assume that raw differential magnitudes, corrected for extinction, have been computed (see Chapter XIII in Photoelectric Photometry of Variable Stars, D. S. Hall and R. M. Genet, Fairborn Observatory, 1982). For each night we have three possible differential magnitudes:

$$\Delta m_{12} = m_1 - m_2 \quad \text{(planet-comparison star 2)}$$

$$\Delta m_{13} = m_1 - m_3 \quad \text{(planet-comparison star 3)}$$

$$\Delta m_{23} = m_2 - m_3 \quad \text{(star 2-star 3)}$$

Δm_{12} and Δm_{13} can be averaged together, if desired, while Δm_{23} serves to establish the overall stability of the photometry, since it should be constant from night to night. Obviously, if star 2 or star 3 turns out to be variable, it can be eliminated from the reductions.

Distance Correction

We must correct the planetary differential magnitudes to a fixed mean heliocentric opposition distance (HP), 5.208 a.u. for the Galilean satellites, 9.539 a.u. for Saturn's satellites,

19.191 a.u. for Uranus, and 30.071 a.u. for Neptune. For this correction, we require the geocentric distance (GD) and heliocentric distance (HD) for the date of observation as given in the Astronomical Almanac. The geocentric distances for each planet are listed for each day in tables located in Section E of the Almanac, under the heading "True Geocentric Distance." The heliocentric distances are given in a different table in Section E, under the heading "Radius Vector." Both quantities must be interpolated to the universal time of the actual observation.

The distance correction, Δd, in magnitudes is given by

$$\Delta d = 5 \{ \log(HP) + \log(HP - 1.) - \log(GD) - \log(HD) \}.$$

Solar Phase Angle Correction

If the coefficients for the solar phase angle correction are known (Table 2-3), then we simple need to look up the phase angle "i" for the date of observation in the Almanac, listed in the tables called "Ephemeris for Physical Observations," also in Section E. Phase angles are inconveniently not given in the Almanac for Uranus and Neptune. Approximate values are given here in Table 2-2.

TABLE 2-2

Solar Phase Angles

Number of days from opposition	Phase Angle Uranus (degrees)	Phase Angle Neptune (degrees)
10	0.55	0.35
20	1.10	0.65
30	1.60	0.95
40	2.05	1.25
50	2.45	1.50
60	2.75	1.65
70	2.95	1.80
80	3.05	1.85
90	3.10	1.90
100	3.00	1.85
110	2.85	1.75
120	2.60	1.60

The equations for phase angle corrections, Δp, in magnitudes are given in Table 2-3 for the V or y magnitudes. For further details, see Lockwood (1977) and Millis and Thompson (1975).

TABLE 2-3

Solar Phase Curve Equations for V or y Magnitudes as a Function of Solar Phase Angle i

Io (near eastern elongation, θ = 70°-115°)	
$\Delta p = 0.0579i - 0.00240i^2$	($0° \leq i \leq 12°$)
$\Delta p = 0.0161i$	($6° \leq i \leq 12°$)
Io (near western elongation, θ = 215°-275°)	
$\Delta p = 0.0473i - 0.00142i^2$	($0° \leq i \leq 12°$)
$\Delta p = 0.0205i$	($6° \leq i \leq 12°$)
Europa (near eastern elongation, θ = 90°-112°)	
$\Delta p = 0.0114i$	($6° \leq i \leq 12°$)
Europa (near western elongation, θ = 200°-250°)	
$\Delta p = 0.0149i$	($6° \leq i \leq 12°$)
(near western elongation, θ = 285°-305°)	
$\Delta p = 0.0164i$	($6° \leq i \leq 12°$)
Ganymede (near eastern elongation, θ = 50°-85°)	
$\Delta p = 0.0445i - 0.00189i^2$	($0° \leq i \leq 12°$)
$\Delta p = 0.0139i$	($6° \leq i \leq 12°$)
Ganymede (near western elongation, θ = 235 - 295°)	
$\Delta p = 0.0542i - 0.00217i^2$	($0° \leq i \leq 12°$)
$\Delta p = 0.0149i$	($6° \leq i \leq 12°$)
(near western elongation, θ = 320°-345°)	
$\Delta p = 0.0262i - 0.00045i^2$	($0° \leq i \leq 12°$)
$\Delta p = 0.0149i$	($6° \leq i \leq 12°$)
Callisto (near eastern elongation, θ = 55°-95°)	
$\Delta p = 0.1014i - 0.00471i^2$	($0° \leq i \leq 12°$)
$\Delta p = 0.0183i$	($6° \leq i \leq 12°$)
Callisto (near western elongation, θ = 220° - 300°)	
$\Delta p = 0.0689i - 0.00268i^2$	($0° \leq i \leq 12°$)
$\Delta p = 0.0183i$	($6° \leq i \leq 12°$)
Titan	
$\Delta p = 0.004i$	($0° \leq i \leq 6°$)
Rhea	
$\Delta p = 0.025i$	($0° \leq i \leq 6°$)
Uranus	
$\Delta p = 0.003i$	($0° \leq i \leq 4°$)
Neptune	
$\Delta p = 0.007i$	($o° \leq i \leq 2°$)

Opposition Magnitude

The observed differential magnitudes Δm_{12} or Δm_{13} are said to be "corrected to opposition" when the effects of changing distance and phase angle are taken out, e.g.:

$$\Delta m_{12}\ \text{(opposition)} = \Delta m_{12}\ \text{(observed)} - \Delta p + \Delta d.$$

Finally, in order to express the magnitude on the UBV or uvby system, we need to add the value for the magnitude of the comparison star used,

$$\Delta m \text{ (planet at opposition)} = m \text{ (comp)} + \Delta m_{12} \text{ (observed)} - \Delta p + \Delta d.$$

Orbital Phase Variations of the Galilean Satellites

After the above corrections have been made, the observer may wish to correct the magnitudes of the Galilean satellites to a common orbital phase angle, near 90° (eastern elongation) or 270° (western elongation). This correction is shown in Table 2-4 for Io, Europa, and Callisto, for the y magnitude, based on Figure 2-2.

TABLE 2-4

Orbital Phase Corrections near Elongation for Three of the Galilean Satellites*

Phase (deg)	Δy for Io	Phase (deg)	Δy for Europa	Phase (deg)	Δy for Callisto
60	-0.025	50	-0.033	230	-0.020
70	-0.015	60	-0.022	240	-0.009
80	-0.005	70	-0.012	250	-0.002
90	-0.005	80	-0.006	260	-0.003
100	0	90	-0.002	270	-0.008
110	-0.005	100	0	280	-0.014
120	-0.005	110	-0.002	290	-0.020
130	-0.010	120	-0.011	300	-0.028
140	-0.015	130	-0.025	310	-0.048

*Ganymede excluded, since we have not observed it. See Millis and Thompson (1975).

The orbital phase curve for Rhea is fairly sinusoidal. We correct the observations to eastern elongation (where Rhea is brightest) using the equation

$$\Delta m = -0.095(1-\sin \theta),$$

where θ is the orbital phase angle, measured from 0° at superior conjunction.

The orbital phase angles are computed from data given in Section F of the Almanac in the table titled "Universal Time of Superior Geocentric Conjunction." The time, t_o, given as date, hour, and minute in the Almanac must be converted to days + fraction. Then, the equation for orbital phase angle is

$$\theta(\text{deg}) = \frac{(t_1 - t_o)}{P} \times 360,$$

where t_1 is the time of observation (days+fraction), and P is the synodic period of the satellite, given in Table 2-5. This equation

TABLE 2-5

Synodic Periods of Satellites

Io	1.7699 days
Europa	3.5541 days
Ganymede	7.1663 days
Callisto	16.754 days
Titan	15.971 days
Rhea	4.5208 days

disregards the Earth's orbital motion but is sufficiently accurate if the time t_o from the Almanac is within a few weeks prior to the time of observation t_o. If an Almanac is unavailable, the time t_o can be read with fair accuracy from the graphical configurations of satellites given in the annual Observer's Handbook published by the Royal Astronomical Society of Canada and in Sky and Telescope magazine. (Note that in the Almanac the times of greatest eastern elongation are given for Rhea, corresponding to $\theta = 90°$, rather than the time of superior geocentric conjunction as was given for the Galilean satellites.

REFERENCES

Harris, D. L. (1961). Photometry and colorimetry of planets and satellites. In Planets and Satellites (G. P. Kuiper and B. M. Middlehurst, eds.), pp. 272-342. University of Chicago Press, Chicago.

Lockwood, G. W. (1977). Secular brightness increases of Titan, Uranus, and Neptune, 1972-1976. Icarus 32, 413-430.

________, and Thompson, D. T. (1979). A relationship between solar activity and planetary albedos. Nature 230, 43-45.

________, Thompson, D. T., and Lumme, K. (1980). A possible detection of solar variability from photometry of Io, Europa, Callisto, and Rhea, 1976-1979. Astronomical Journal 85, 961-968.

Millis, R. L., and Thompson, D. T. (1975). UBV photometry of the Galilean satellites. Icarus 26, 408-419.

Morrison, D., Morrison, N. D., Lazarewicz, A. R. (1974). Four-color photometry of the Galilean satellites. Icarus 23, 399-416.

Noland, M., Veverka, J., Morrison, D., Cruikshank, D. P., Lazarewicz, A. R., Morrison, N. D., Elliot, J. L., Goguen, J., and Burns, J. A. (1974). Six-color photometry of Iapetus, Titan, Rhea, Dione, and Tethys. Icarus 23, 334-354.

3. PHOTOMETRY OF COMETS

Michael F. A'Hearn

I. INTRODUCTION

Before discussing the photometry of comets, it is desirable to review briefly the nature of comets. In addition to reminding the reader of cometary terminology, this brief review will also emphasize that photometry is "useful" only insofar as it helps us to understand the nature of comets. The most widely accepted picture of a comet is that originated by Whipple (1950). An observable comet is hypothesized to develop from a solid nucleus, which has a size of order 1 km and which consists of various ices plus small, refractory particles, and perhaps a large, rocky core. There is no direct evidence for a rocky core and it is even a matter of dispute among astronomers whether anyone has ever seen the nucleus at all. There are other models, which do not postulate a solid nucleus (e.g. Lyttleton, 1951) but they also must produce the same observed phenomena which are to be measured by photoelectric photometry; thus our discussions of observations will be applicable to any model of comets. With rare exceptions (e.g., P/Schwassmann-Wachmann 1 in a nearly circular orbit and Bowell 1980b which was perturbed by Jupiter into a hyperbolic orbit), these cometry nuclei travel in very elongated orbits with orbital periods ranging from 3.3 years (P/Encke) to several million years which is the longest period a comet can have and remain gravitationally bound to the sun.

As a cometary nucleus approaches the Sun, it gradually warms up and the ices begin to vaporize. These vaporized gases expand into the vacuum of space and drag with them the small, refractory grains which are commonly referred to as dust. Some of the molecules are ionized or dissociated by sunlight, some may be ionized by other means, and some participate in chemical reactions with other molecules. This cloud of gas and dust forms the observable coma or head of the comet and has a size of order 10^5 to 10^6 km. Meanwhile the dust and the ions are driven away from the Sun by radiation pressure and by a magnetic interaction with the solar wind, respectively, to form the dust tail and the ion tail. Different photometric techniques are needed to study these different parts of comets.

According to the picture developed originally by Oort (1950), there is a large reservoir of comets orbiting the Sun at large distances extending to perhaps 30,000 to 50,000 AU. Most of these comets are in orbits of sufficiently low eccentricity that they never enter the planetary region. Their orbits are occasionally perturbed by stars which pass near the Sun and some comets thus enter the planetary region on highly elongated orbits and become observable. Perturbations by the planets at successive approaches to the Sun will transform this orbit in a random-walk fashion until either the comet is ejected from the solar system on a hyperbolic orbit or the comet enters a short period (P $\leq$ 200 years) orbit in which it rapidly (perhaps in a few thousand orbital periods) loses all its volatile material, a typical loss rate being a few meters of icy material per perihelion passage.

A typical photoelectric tracing of the spectrum of a comet is shown in Figure 3-1. The continuous spectrum is due to sunlight scattered by the dust while the emission features are due to fluorescence by the various species indicated. In the fluorescence

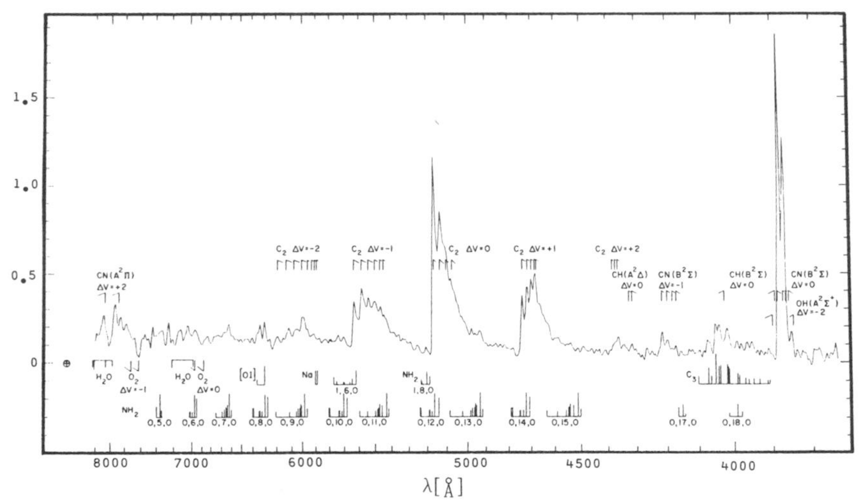

Figure 3-1. Spectral scan of Comet Kohoutek 1973XII. All common emission features of neutral species are indicated schematically even though some are not visible in the spectrum. Relative strengths of different features vary with heliocentric distance of the comet and with size of the diaphragm. See A'Hearn (1975) for details regarding spectrum.

process, the atom or molecule absorbs a photon of sunlight raising the atom (molecule) to an excited state. The atom (molecule) then spontaneously decays back toward the ground state in one or several steps, emitting a photon at each step. Note that none of the identified species are likely to be present in the nucleus; rather they are dissociation products and/or reaction products of the species thought to be present in the nucleus. When considering cometary photometry, these aspects of the spectrum must be kept in mind. Different features in the spectrum typically appear at different heliocentric distances but otherwise cometary spectra are remarkably similar to one another. The only major spectral differences from one comet to another are in the ratio of emission band strength to continuum (related to the gas-to-dust ratio) and in the relative strengths of features due to ions.

II. GOALS AND RESULTS

Broadband Magnitudes

As in most other branches of astronomy, cometary photometry began with very broad bandpasses, typically the full visual or photographic range. It has long been common practice to distinguish between the integrated magnitude, m_1, referring to the total light from all parts of the comet and the so-called nuclear magnitude, m_2, referring to the light from the bright, central condensation which is rarely, if ever, that of the true nucleus. Values of m_2 are typically obtained photographically, often from astrometric plates for which the exposure is as short as possible and intended to show only the nuclear region. Values of m_1 are typically obtained visually using small (10- to 20-inch) telescopes although they are also obtained from long-exposure photographs.

The variations of m_1 and m_2 with heliocentric distance (after correcting for the geocentric distance using an inverse square law) have been very valuable in delineating the overall behavior of comets. For example, there are some comets (e.g., P/Arend-Rigaux and P/Neujmin 1) for which m_2 exhibits an inverse square law behavior at large heliocentric distances, viz $m_2 = M_2 + 5 \log r + 5 \log \Delta$, where r and Δ are the heliocentric and geocentric distances of the comet (in AU) and M_2 is the comet's absolute magnitude in the relevant bandpass. This behavior has been used to infer that the true nucleus is being observed since, if there were activity of any kind in the comet, one would expect that activity to decrease as the heliocentric distance increases thus varying the effective cross section of the comet which in turn would lead to deviations from the inverse square law. The two cited comets also exhibit a phase dependence of the type exhibited by asteroids (Sekanina, 1976). For a variety of reasons, the interpretation is far from universally accepted, even by some who have pointed out the behavior, but the behavior is certainly suggestive of minimal activity.

The magnitude, m_1, on the other hand, virtually never exhibits an inverse square law behavior with heliocentric distance (it is, of course, not quoted for a comet which appears stellar, i.e. when a coma is not visible). Since the visual bandpass contains both emission bands (primarily of C_2) and continuum, this magnitude is a useful indicator of the total amount of material, gas and dust, in the coma. Its variation with heliocentric distance can therefore be used to infer the vaporization properties of the nuclear ices. In this connection, for example, Whipple (1978) has recently shown that, for comets which are thought on dynamical grounds to be entering the inner solar system from the Oort cloud, m_1 exhibits a significantly different behavior with heliocentric distance (varying much more slowly) than for other comets or even for these same comets as they leave the inner solar system. Whipple has interpreted this difference as a sign of increased volatility of the outer layers of these comets produced by the bombardment of cosmic rays over the 5-billion year age of the solar system.

In stellar photometry, visual and photographic magnitudes have been entirely supplanted by photoelectric photometry in the well-known UBV system as well as in several other broadband systems. The UBV system and other similar systems are particularly useful for stars because they provide colors which are diagnostic of physical conditions in stars. Although some attempts have been made to apply the UBV system to comets, these have not been very profitable because the colors, U-B and B-V, are not at all diagnostic of any physical condition in comets.

A single one of these bands, say either B or V, might be useful for obtaining values analogous to either m_1 or m_2 but, with some exceptions, this is not expected to be a profitable endeavor. Estimates of m_1 are simply not practical with photoelectric equipment (except perhaps with imaging detectors such as CCD arrays) because photoelectric photometry necessitates use of a certain field of view on the sky which, in general, cannot be made to include the whole comet. The extra-focal methods employed by visual observers are moderately effective but not readily adapted to photoelectric instruments. The primary limitation to the accuracy of estimates of m_1 is not in the precision of the detector but rather in deciding where the comet stops and the sky background begins. Thus the increased precision of photoelectric measurements would not lead to increased accuracy and, in any case, the precision of visual estimates seems sufficient for dealing with the physically interesting questions about vaporization rates.

One useful application of broadband photometry is in obtaining values of m_2 for distant comets which show little activity. When comets are near the Sun and active, the difficulty in separating the brightness of the nuclear region from that of the coma makes the improved precision of photometric techniques pointless. For distant comets, however, the nuclear region dominates the total brightness. In this case, broadband photometry can be very useful in determining the variation of brightness with heliocentric distance and testing whether the brightness is truly following an inverse square law. It is also true that very distant comets usually show a purely continuous spectrum. In this case, comparison of broadband photometry of comets with that of solar-type stars can determine the color of the cometary material. This technique has been applied, for example, to P/Tempel 2 (Zellner et al. 1979) and led to the then surprising result that this comet underwent outbursts at r = 3 AU; based on these and other data we are now coming to the view that the occurrence of outbursts is widespread rather than limited to a few peculiar comets. However, most of the situations in which broadband photometry would be useful involve very faint comets, often fainter than 18^m; thus the programs are limited to very large telescopes.

Another area in which broadband photometry might be useful is the determination of rotation periods of cometary nuclei. Most determinations of rotation periods have been obtained from photographs of jets and halos but in principle one should see brightness variations due to rotation as are seen for most asteroids. This can be done only if one can isolate the nuclear region very well, either by picking a very distant (thus very faint) comet or by picking a comet close to earth with relatively little dust. The only published result based on this technique is the five-hour period for comet P/d'Arrest determined by Fay and Wisniewski (1978). These measurements could be done most easily with narrowband filters which isolate the continuum but broadband measurements can be used.

Narrowband Photometry

Most photoelectric photometry of comets is now done with narrowband interference filters chosen specifically to isolate features of interest in cometary spectra. The use of such filters dates back at least to 1954 (Schmidt and van Woerden, 1957) and is now widespread. The filters are usually chosen to isolate single emission features and/or a portion of the continuum containing no emission features. With a suitable set of filters, a wide variety of programs can be pursued. For convenience we will separate them into studies of the dust (continuum) and studies of the gas (emission lines and bands) but generally the two are intertwined. Many of the examples will be taken from our own work with various collaborators, particularly with R. L. Millis, because these data are most readily available to us; these are, however, just examples and do not represent a comprehensive survey of photometric studies of comets.

The filters which we have been using for the last several years are shown in Figure 3-2 where they have been superimposed on the spectrum of comet P/Tuttle. (The spectrum was synthesized by S. Larson using data from different sources for different wavelength intervals.) The filters isolate the strongest emission features of OH, NH, CN, C_3, and C_2 as well as continuum bands at λ 3300, λ 3675, and λ 5240. In the past, we have ourselves used other filters (e.g., for the $\Delta v = +1$ sequence of C_2 with head at

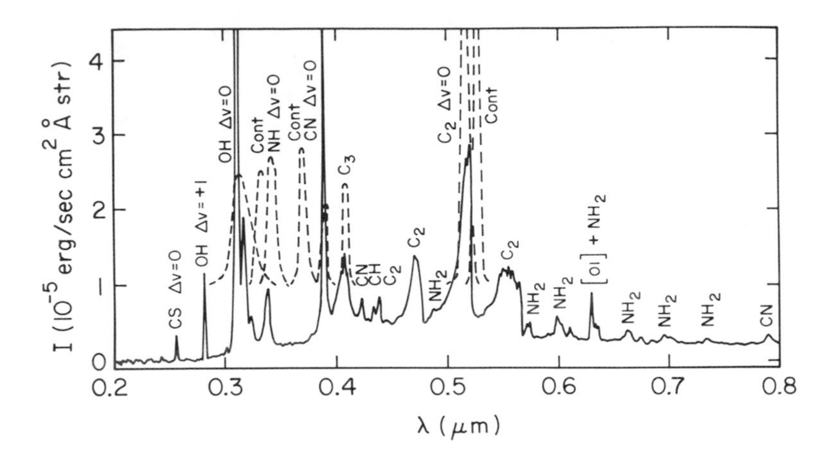

Figure 3-2. Typical filters for cometary photometry. The profiles of the filters used for cometary photometry over the past several years by A'Hearn and Millis are superimposed on a spectral scan of Comet P/Tuttle 1980h which was synthesized from various data by S. Larson and J. R. Johnson (private communication).

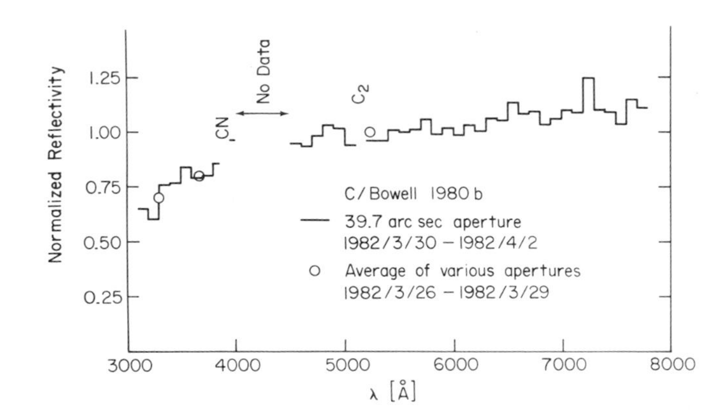

Figure 3-3. The color of Comet Bowell 1980b. Spectrophotometry by Schleicher and A'Hearn provides the variation of reflectivity with wavelength. The open circles were obtained from the color excess of the comet relative to the solar analogs as observed with the continuum filters shown in Figure 3-2.

λ 4737 and for continuum bandpasses at λ 3920 and λ 4520. Note that most molecular emission features are bands, i.e., groups of lines, which are often denoted by the change in a vibrational quantum number, Δ v.) and other investigators have used a variety of other filters, some similar to ours and others different. In the near future, we will switch systems to use the IAU sponsored standard system discussed in Section III below.

Finally, we note that the continuum bandpasses are best for studying nuclear rotation as discussed above (Section II).

Continuum Studies

Perhaps the most obvious project in studying the continuum is to determine the colors of the grains. Figure 3-3 shows the normalized reflectivity of the grains in Comet Bowell (1980b) as determined from spectral scans and superimposed on this we have plotted the results of filter photometry. This comet is an unusual one in having virtually no emission features so that even broadband photometry would give useful colors. The only data plotted here, however, are from the filters normally used to isolate the continuum. It is clear that filter photometry, based solely on the color difference between the comet and stars like the sun converted to linear units, yields an accurate measure of the color of the cometary grains, which might be characterized as slightly pink in this particular case. Observations of many other comets indicate variations in the color from gray (a horizontal line in Figure 3-3) to twice as red as Bowell (a line twice as steep as in Figure 3-3). There are not yet, however, sufficient data to decide whether the differences are due mainly to differences in the scattering geometry or to differences in particle size and/or composition. Our own data from filter photometry also show that the continuum appears bluer in comets with high gas-to-dust ratios but it is possible that this is an artifact due to increasing contamination of the continuum bandpasses by weak emission lines.

Another interesting project is to determine the variation in brightness with scattering angle as the sun-comet-earth geometry changes. This type of study is more difficult because one must first separate temporal changes in the amount of dust from geometrical changes in the brightness of a given amount of dust. Furthermore, one must correct for the different volumes subtended at the comet by the photometer diaphragm. If the comet is "well behaved" in the sense of maintaining proportionality between gas and dust, then the temporal changes can be monitored by observing the amount of gas present. Figure 3-4 illustrates the back-scattering function of the grains in Comet P/Stephan-Oterma (Millis *et al*. 1982). Similar results have been obtained for other comets by Dobrovol'skij's group at Dushanbe. Measurements such as these can be compared with laboratory measurements to determine the types of particles that could be present.

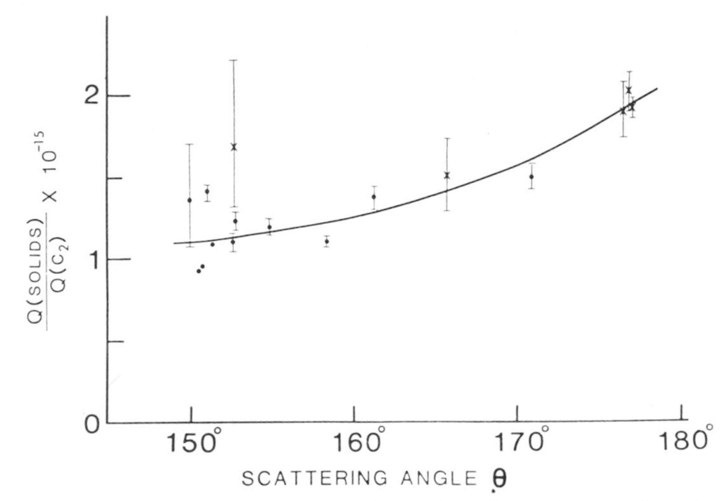

Figure 3-4. The back-scattering function for the grains in Comet P/Stephan-Oterma 1980g. The abscissa is the scattering angle (180°-phase). The plotted ordinate is the ratio of dust production to gas production assuming no variation of scattering efficiency (approximately the ratio of continuum flux to emission band flux for a fixed aperture) but we infer that the change in the ratio is due to a change in scattering efficiency with angle rather than a real change in relative abundance. See Millis et al. (1982) for data used in this figure.

Another program which might well be carried out via photometry is a study of the variation of continuum color with position, e.g., with projected distance from the nucleus. Interesting results using spectrophotometric techniques have been reported for Comet P/Schwassmann-Wachmann 1 by Cochran et al. (1982) who found, during an outburst, a systematic variation of color with distance from the nucleus. They interpreted this in terms of a change of particle size with time during outburst. Similar studies could profitably be carried out using filter photometry on nearer comets to search for such differences indicating a segregation of particles by size and/or type.

Studies of Emission Features

One of the earliest goals of photoelectric photometry of comets was simply to determine the absolute abundance of various gaseous species by measuring the absolute flux in certain emission bands. This is still the basis of most studies today although most do require the use of additional modeling for interpretation. In our own work, for example, we apply corrections for diaphragm size by using a Haser model and convert total abundances into production rates using an assumed lifetime for the observed species. Nevertheless, qualitatively similar conclusions can often be reached by a direct study of the fluxes of the emission features.

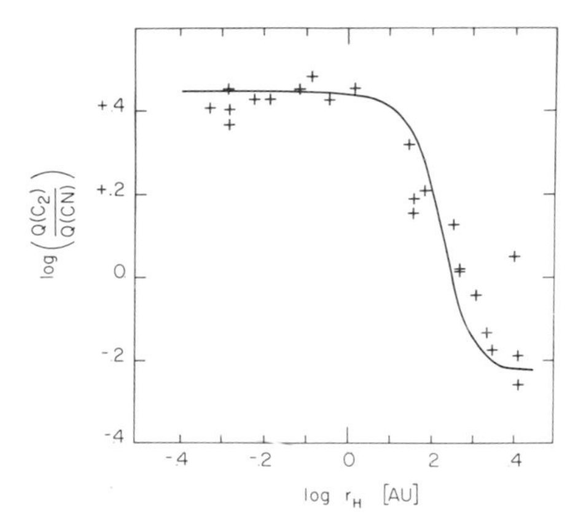

Figure 3-5. Relative production rates in Comet West 1976VI. Apparently the production of C_2 relative to that of CN decreased markedly when the comet receded past 2 AU. This effect would show up equally well in the ratio of emission band fluxes provided they were measured with a large aperture of fixed size projected at the comet. See A'Hearn and Millis (1980) and A'Hearn et al. (1977) for the original data.

One possible study is of the change in relative abundances (or production rates) for a particular comet as a function of heliocentric distance. This can elucidate the source of different observed species by comparison with chemical models of the coma. Figure 3-5 shows the change in the ratio of C_2 and CN production rates in Comet West (1976VI) as the heliocentric distance increased (cf. A'Hearn, 1981). This type of change was interpreted to indicate that the parent of C_2 was trapped in H_2O ice which ceased vaporizing near r = 2 AU. It might also be interpretable in terms of changes in the rates of the chemical reactions that produce C_2. A more recent result, shown in Figure 3-6, is the OH production of Comet Bowell (1980b). This comet apparently had a large halo of grains released by a very volatile ice at large heliocentric distance. Near 4 AU, the H_2O in or on the grains began to evaporate and by the time the comet reached perihelion, most of this ice had disappeared (A'Hearn et al. 1983).

A different type of study involves comparisons of chemical compositions among many comets. This type of study can easily be done directly with emission band fluxes if one is careful to always observe the same physical volume in the different comets. Studies of C_2, CN, and C_3 indicate very little variation from one comet to another (e.g. A'Hearn 1981). There are systematic variations with heliocentric distance but not with age of the comet as shown, e.g., in Figure 3-7. This in turn suggests that comets are not stratified (otherwise old comets would have a different composition from new comets) and that most comets formed in a homogeneous region of the solar nebula. There are a few exceptions to this uniformity

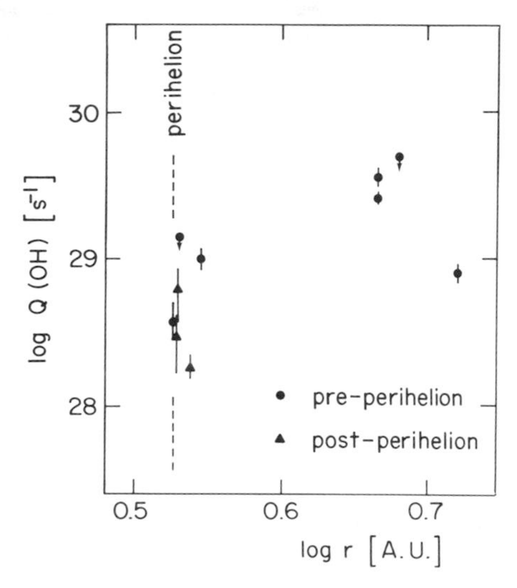

Figure 3-6. Production of OH in Comet Bowell 1980b. As this comet approached the Sun, the OH emission was initially undetectable but it appeared and rose rapidly as the comet passed 4 AU. By the time the comet reached perihelion at 3.3 AU the OH emission had decreased dramatically. Two data points shown as upper limits (arrows pointing downward) represent 2 standard deviations. Data for this figure will be published elsewhere.

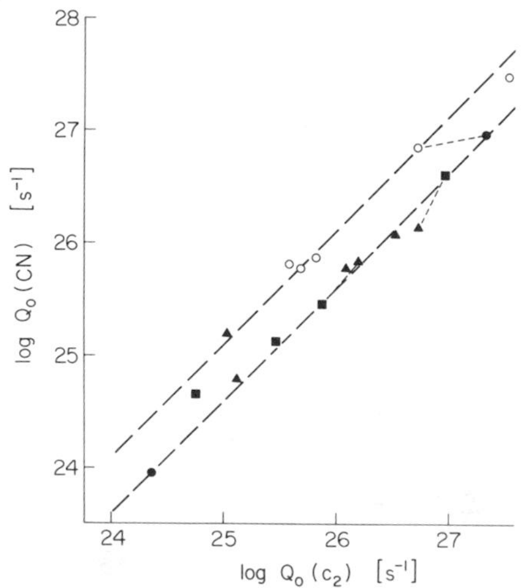

Figure 3-7. Correlation between C_2 and CN production. Comets observed at large heliocentric distance ($>$ 2 AU) exhibit a smaller ratio of C_2 to CN than do comets observed at small heliocentric distance ($<$ 1.5 AU). There is no other systematic variation observed as a function of dynamical age of the comet, gas-to-dust ratio, or other likely parameters. This strong correlation could be shown in the emission band fluxes provided all data were taken through the same diaphragm. Figure based on data presented by A'Hearn and Millis (1980).

but they are not yet understood. More recent studies using photometry of OH (A'Hearn and Millis 1983) suggest a similar uniformity in relative abundances from one comet to another. This entire area of study requires much more work.

It also turns out that many comets show unusual behavior with heliocentric distance, a prime example being Comet Encke. This comet exhibits a dramatic decrease in the production of OH when it gets to within about 0.7 AU of the Sun. From there through perihelion, it appears to be much fainter in OH than one would predict from the brightness before reaching 0.7 AU (Thompson et al. 1982). Activity such as this can be discovered only by continuous monitoring of comets, a practice most easily carried out with filter photometry.

These examples serve primarily to indicate the types of programs that can be carried out. Most of them do require systematic observations over a long period of time; they cannot be carried out in a single, week-long observing session. For this reason, many of them are best carried out at smaller facilities where telescope time can be obtained more easily than at major national observatories. The equipment required for such programs is discussed in the following section.

III. INSTRUMENTATION

The instrumentation for photoelectric photometry of comets requires some special consideration. Most standard designs for photometers (see, e.g., Hall and Genet, 1981 or chapter five in this book) are suitable provided that a few points are considered in the implementation of a standard design. Great care must be given to diaphragm design and to choice of filters while a little extra thought must go into the choice of phototube and Fabry lens. We will discuss only these deviations from standard designs.

Diaphragms

Because comets are invariably larger in angular size than the projected size of any diaphragm likely to be used, the diaphragm plays quite a different role in cometary photometry than it does in most other photometry. Because the diaphragm actually defines the portion of the comet being observed, it is first of all essential to know precisely the size and shape of the diaphragm being used. (This is not necessary in photometry of stars or asteroids because in those cases the diaphragm serves only to minimize the sky contribution and the entire object of interest is contained within the diaphragm.)

There are also other considerations regarding the choice of diaphragm which depend on the goals and interests of the observers. To understand these considerations, it is necessary to review the distribution of light in the cometary coma. In the ideal comet, the continuum (sunlight scattered from the solid particles) has a surface brightness which varies as $1/\rho$, where ρ is the projected distance from the nucleus. Most comets, of course, show some form of azimuthal asymmetry, some show expanding halos during outbursts, and some show sharp edges, but the main point is that in nearly all cases the brightness of the continuum is sharply peaked near the nucleus. The gases, however, show quite different distributions. A few species, notably [OI] and C_3, are also concentrated very strongly toward the nuclear region although the spatial profiles do not exhibit the simple $1/\rho$ dependence which is exhibited by profiles in the continuum. The gases which dominate the optical spectrum, OH, CN, and C_2, are all produced at distances greater than 10^4 km from the nucleus and are then destroyed at distances greater than 10^5 km from the nucleus. As a result, the surface brightness in an emission band of one of these species varies much more slowly than $1/\rho$ near the nucleus and drops off sharply only at projected distances, ρ, greater than 10^5 km from the nucleus. A natural consequence of the difference in behavior is that the emission-to-continuum ratio, often taken as a measure of the gas-to-dust ratio in comets, increases systematically as the size of the diaphragm increases. Thus, an observer interested primarily in the color of the continuum would obtain the least contamination from emission features by using a small diaphragm. Conversely, an observer interested primarily in gaseous emission would have the least contamination from continuum with a very large diaphragm. Most observers, of course, will want to use both large and small diaphragms.

One of the interesting problems in cometary photometry is the determination of the spatial distribution of various species. It would be best to use long slit spectroscopy to get the purest spectral isolation or imaging devices to obtain two-dimensional information. Nevertheless, one can study spatial distribution with conventional photometers. To do this, the best approach is probably the use of annular diaphragms rather than simple, circular ones. To study spatial distributions with conventional diaphragms, it is necessary to difference successive observations, a procedure which always enhances any noise.

Since the varying geocentric distances of comets lead to differences in image scale at the telescope focal plane, the fixed diaphragms found in most photometers will correspond to varying linear sizes at the comets. For typical geocentric distances near 1 AU, the linear scales cited above (10^4 km and 10^5 km) correspond to angular radii near 10 arcsec and 2 arcmin. It should be emphasized that diaphragms of several arcminutes, although rarely found in standard photometers, are essential for cometary photometry. Unfortunately, one cannot simply replace the small diaphragms with large ones. These large diaphragms, when filled with a source, can lead to a beam of light which is too large for the Fabry lens, and a careful calculation of the beam size should be made before adding larger diaphragms. When designing a photometer with conventional diaphragms, a good set of diaphragms might have diameters of 3 arcsec, 10 arcsec, 30 arcsec, 100 arcsec, and 300 arcsec. More diaphragms, which are more closely spaced in size, would certainly be better but might also become unwieldy. For use on smaller telescopes, the 3 and 10 arcsec diaphragms might be impractical but one might also add a diaphragm of 1000 arcsec, a size which would be impossible on most large telescopes.

Because of the variation in linear scales for comets at different geocentric distances, data from different comets (or even from the same comet at different heliocentric distances) cannot be directly compared without mathematical extrapolations of sometimes dubious validity. To eliminate this problem, K. Jockers has suggested (private communications) that one use a calibrated iris diaphragm. One could then calculate, for each comet on each night of observation, the angular diameters corresponding to physically interesting, linear projected radii (e.g., 10^4 km, 3×10^4 km, 10^5 km, 3×10^5 km) and adjust the diaphragm precisely to those sizes. Although the iris might not provide a perfectly circular diaphragm, a good iris should be sufficiently close to circular and one could, for example, calibrate the iris opening in terms of the radius of a circle of equal area. This technique appears very promising although we have not yet seen it implemented. If implemented properly, it would permit direct comparison of fluxes measured for different comets at different times.

Filters

As emphasized in Section II, in order to carry out useful photoelectric photometry of comets, it is almost essential to use interference filters designed to isolate spectral features of interest, both emission features and portions of the continuum. The choice of which features to measure depends primarily on the observer's interests, but there are some practical limitations.

The OH emission band at 3085 Å is of great interest, but the atmospheric attenuation is so great that observers at sea level would be well advised to ignore it entirely. Even at 7000 feet (Lowell Observatory, Flagstaff, Ariz.), the extinction coefficient at this wavelength is typically 1.25 to 1.50 magnitudes per air mass, necessitating a very careful treatment of extinction. When observing comets in large apertures, the sky brightness also becomes appreciable and the natural emissions of the night sky (as well as any local man-made emissions) must be taken into account. Figure 3-8 shows a low resolution plot of a typical atmospheric extinction curve (based on data from Lowell Observatory) and a typical night sky brightness curve (adapted from Roach and Gordon, 1973), and we have indicated the locations of numerous cometary emission features. It is clear that both the shortest and the longest wavelengths suffer from high sky-brightness. In the far red, this emission is due, in large part, to OH vibrational emission features which vary greatly with time and with position in the sky. These considerations are particularly important in those few cases where a particular species has several emission features and one can choose which feature to observe based on contrast with the sky after atmospheric

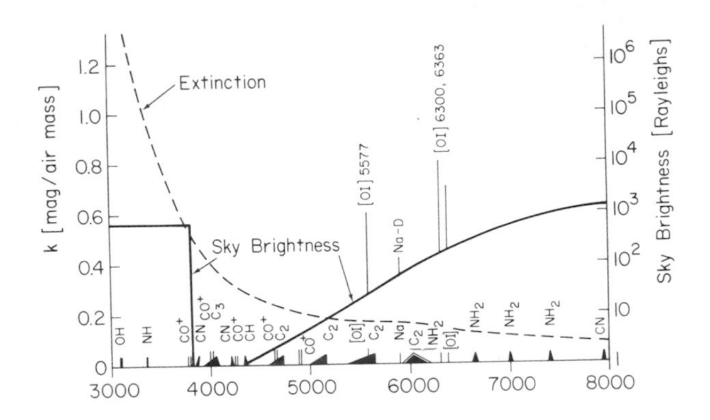

Figure 3-8. Observing conditions as a function of wavelength. The emission features in cometary spectra are shown schematically along the bottom of the figure (see Figures 3-1 and 3-2 for more details). The two curves show the atmospheric extinction at Flagstaff, Arizona (unpublished data from A'Hearn and Zipoy) and the typical brightness of the night sky from natural sources (from Roach and Gordon, 1973). Man-made sources of sky brightness should be added but depend critically on the observer's location.

extinction. In other cases, however, one must simply live with a large extinction coefficient and/or large sky corrections. Another practical limit lies in the strength of the features of interest relative to other, unrelated features. Weak emission features are difficult to separate either from the continuum or from nearby, stronger emission features. The continuum, itself, is difficult to separate from emission bands except in a few wavelength regions and in very dusty comets with almost no gas.

With the above comments in mind and allowing for the sensitivity of typical photomultipliers, most early cometary photometry emphasized the $\Delta v = 0$ sequence (i.e. all the lines corresponding to no change, $\Delta v = 0$, in the vibrational quantum number, v) of CN extending to the blue from 3883 Å and either the $\Delta v = +1$ or the $\Delta v = 0$ sequence of C_2 extending to the blue from 4737 Å and 5165 Å, respectively. More recent work has included C_3 (peaked near 4050 Å), NH (centered near 3365 Å) and, because of its importance, OH (centered at 3085 Å). Some photometry has been carried out on the Na-D lines (5889, 5895 Å) but the observations are difficult because the night sky has emission in these lines which gets particularly strong near twilight. Other features, such as NH_2, CH, CO^+, H_2O^+, etc. have usually been studied spectroscopically; however when CO^+ is strong, it could be easily studied by photometry.

In order to standardize the photometry and to reduce the costs of the filters themselves, Commission 15 of the IAU, in 1978, established a Working Group to define some standard filters, purchase them, and distribute them to interested astronomers. Thus far, the group has decided on a set of eight filters for the most commonly studied features in comets and the characteristics are given in Table 3-1. Many of these filters have already been distributed to astronomers around the world. Anyone desiring details regarding the distribution of these filters should contact this author. Other filters will also be selected by this group.

When specifying filters for cometary photometry, it is necessary to study cometary spectra carefully and keep several facts in mind. First of all, there is a trade-off between collecting more photons by making the filter bandpass wider and most clearly separating emission features from the continuum by making the filter narrower. For example, in the case of C_3, one must choose between measuring just the strongest peaks in the band near $\lambda\lambda$4020, 4050, and 4070 or measuring the entire band including weak features which extends roughly from 3940 to 4110 Å. Except in comets with very little dust, widening the bandpass to include the weak features has the primary effect of adding more continuum (which must then be subtracted away) so that the emission band represents a smaller fraction of the total light, thus actually decreasing the final signal-to-noise ratio. Furthermore, it allows other emission features to enter the bandpass, specifically one of the bands of CO^+. These trade-offs must be studied separately for every feature to be isolated.

Another point which must be considered is the shape of the bandpass and the way in which the response will shift as the temperature of the filter changes. In general, interference filters will have their response shifted by one- or two-tenths of an Ångstrom for each °C change of temperature. The range of temperatures at which the filter will be used might be from -20°C to +20°C (or even wider in some cases) implying bandpass shifts of several Ångstroms. In order that the response to a sharply-edged emission feature not change with temperature, the response of the filter must remain flat over the intense portion of the feature with an extra allowance for temperature changes. This usually requires an interference filter with at least three interference cavities. Fortunately, making the top of the bandpass square (instead of bell shaped) also narrows the wide skirt of the bell-shaped curve, thus minimizing contamination from other features at nearby wavelengths As can be seen from Table 3-1 and actual cometary spectra, the filters for emission bands were specified as 4 cavity (even more square than 3 cavity) and with flat ($>$ 80% of peak) tops covering all intense parts of the emission bands with an extra 5 Å to spare on the long wavelength side to allow for the shift to shorter wavelengths at temperatures below room temperature. Furthermore, because of the pronounced asymmetries of the CN and C_2 bands (sharp edges on the long wavelength side), the filters for those features have their position specified not by the wavelength at the center but by the wavelength at the long wavelength cutoff. This approach

TABLE 3-1

Comet Filter Specifications

1.	C_2	d-a	4 cavity, nominal center at $\lambda5139$
		$\Delta v = 0$	λ at $T = 0.80\ T_{max}$ (long λ side) : $5175\ ^{+10}_{-5}$
			FWHM = $\Delta\lambda_{.5}$: 90 ± 10
			$\Delta\lambda_{.8} \geq 0.80 \times \Delta\lambda_{.5}$
			$\Delta\lambda_{.01} \leq 1.75 \times \Delta\lambda_{.5}$
			$T_{max} \geq .50$
2.	Continuum		3 cavity, center at $\lambda4845 \pm 10$
			$\Delta\lambda_{.5} = 65 \pm 10$
			$\Delta\lambda_{.8} \geq .75 \times \Delta\lambda_{.5}$
			$\Delta\lambda_{.1} \leq 1.50 \times \Delta\lambda_{.5}$
			$\Delta\lambda_{.01} \leq 2.20 \times \Delta\lambda_{.5}$
			$T_{max} \geq 0.50$
3.	C_3	A-X	4 cavity, center at $\lambda4060 \pm 5$
		$\lambda4050$	$\Delta\lambda_{.5} = 70\ ^{+10}_{-5}$
			$\Delta\lambda_{.8} \geq .77 \times \Delta\lambda_{.5}$
			$\Delta\lambda_{.01} \leq 1.70 \times \Delta\lambda_{.5}$
			$T \leq .5T_{max}$ at $\lambda \leq 4020$ (CO^+ contamination)
			$T_{max} \geq .40$
4.	CN	B-X	3 cavity, nominal center at $\lambda3871$
		$\Delta v = 0$	λ at $T = 0.80\ T_{max}$ (long λ side) : 3890 ± 5
			$\Delta\lambda_{.5} = 50\ ^{+0}_{-10}$
			$\Delta\lambda_{.8} \geq .75 \times \Delta\lambda_{.5}$
			$\Delta\lambda_{.01} \leq 2.20 \times \Delta\lambda_{.5}$
			$T_{max} \geq .20$
5.	Continuum		3 cavity, center at $\lambda3650 \pm 10$
			$\Delta\lambda = 80\ ^{+10}_{-5}$
			$T < .5 \times T_{max}$ for $\lambda \leq 3600$
			$T < .1 \times T_{max}$ for $\lambda \leq 3580$
			$T < 0.5 \times T_{max}$ for $\lambda \geq 3695$
			$T < 0.002 \times T_{max}$ for $\lambda \geq 3860$
			$T_{max} > .25$
			Secondary leak at $\lambda3180$:
			$T_{leak} < 10^{-3}\ T_{max}$ $\Delta\lambda_{leak} < 100$ Å
6.	H_2O^+		Nominal Central Wavelength: 7000Å ± 15Å
			Nominal Half-power width: 175Å ± 17Å
			$T_{peak} \geq 70\%$; $T > 0.80\ T_{peak}$ for $6925 \leq \lambda \geq 7075$Å
			$T < 0.10\ T_{peak}$ for $\lambda \leq 6890$Å; $T < 0.10\ T_{peak}$ for $\lambda \geq 7165$Å
			$\Delta\lambda_{.01} \leq 330$Å $\Delta\lambda_{.001} \leq 450$Å $\frac{d\lambda_o}{dT} = .22$Å/°C
7.	Continuum		Nominal Central Wavelength: 6840Å ± 10Å
			Nominal Half-power width: 90Å ± 9Å
			$T < 0.10\ T_{peak}$ for $\lambda \leq 6775$Å; $T < 0.10\ T_{peak}$ for $\lambda \geq 6905$Å
			$T_{avg} > 50\%$ over entire $\Delta\lambda_{.5}$
			$\Delta\lambda_{.01} \leq 2.2\ \Delta\lambda_{.5}$; $\frac{d\lambda_o}{dT} = .22$Å/°C
8.	CO^+		Nominal Central Wavelength: 4260Å ± 5Å
			Nominal Half-power width: 65Å ± 5Å
			$T_{peak} \geq 40\%$ $T > 0.80\ T_{peak}$ for $4235 \leq \lambda \leq 4285$Å
			$T < 0.10\ T_{peak}$ for $\lambda \leq 4215$Å; $T < 0.10\ T_{peak}$ for $\lambda \geq 4300$Å
			$\Delta\lambda_{.01} \leq 120$Å $\Delta\lambda_{.001} \leq 175$Å; $\frac{d\lambda_o}{dT} < .20$Å/°C

allows a wider tolerance on the bandwidth than would be possible by specifying the center of the bandpass, thereby making the filters slightly less expensive for a given astronomical quality.

A final note of caution regards the blocking of the filters, i.e. the transmission at wavelengths far outside the bandpass. Since pure interference filters usually have several bandpasses at different wavelengths, the filter requires special blocking to eliminate the extraneous ones at all wavelengths to which the phototube is sensitive. This is particularly important for narrow bandpass filters such as those being considered here. For most situations, a convenient rule of thumb is that the blocking transmission multiplied by the bandwidth of the phototube (e.g., 10^{-5} x 5000 Å = .05 Å) should be negligible compared to the peak transmission multiplied by the bandwidth of the filter (e.g., .40 x 80 Å = 32 Å). In some cases, e.g., the case of measuring a weak feature in the presence of numerous stronger ones, even more stringent blocking may be required. The ultimate goal, of course, is to ensure that light from outside the bandpass does not contribute to the measured signal.

Fabry Lens and Phototube

There are two considerations in choosing the phototube: quantum efficiency and spectral range. The widely used 1P21 (or cheaper 931A) by RCA is quite suitable for measuring the strong bands of C_2 and CN as well as continuum and other emissions (C_3, NH, CO^+) in that spectral range. This tube, on the other hand, has the disadvantage of being a side-window type with an oblique cathode. With such a tube, a very fast Fabry lens will often not work well and in view of the large diaphragms used for cometary photometry an end-window tube might be preferred. A tube with a GaAS photocathode (e.g., the C31034 manufactured by RCA) offers a wider spectral range while still maintaining low dark current and high quantum efficiency, albeit at substantially higher price. This gives a much longer baseline in wavelength for studying the color of the continuum. It also brings into reach the [OI] lines at 6300, 6363 Å and several additional bands of NH_2. A tube with near infrared response, even more expensive, also brings in the red system (A-X transitions) of CN.

At the other end of the spectrum, if one wants to study NH or OH, it is essential to specify that both the tube envelope and the Fabry lens be made of fused quartz. This adds substantially to the cost and has the technical disadvantage of allowing larger pulses of background due to cosmic rays. However, the use of fused quartz is essential if one expects to observe features shortward of 3500 Å. As noted above, however, these attempts should probably be limited to observers at moderately high altitudes.

Finally we note that in choosing the Fabry lens, it is essential to choose one with a sufficiently large aperture to allow use of the large diaphragms discussed in Section III above.

Conclusions

-Allow for a variety of calibrated, large diaphragms or a calibrated iris.

-Use interference filters specifically designed for cometary studies.

-Choose the phototube and Fabry lens to match items 1 and 2.

IV. OBSERVING TECHNIQUES

As in Section III, we will discuss here only those aspects in which photometry of comets presents special problems not normally encountered in other types of photometry. Many considerations, such as choice of integration time, proportions of observing time on comet and sky, etc. are no different for cometary photometry than for stellar or asteroidal photometry.

Acquisition and Tracking

Finding faint comets is largely an art which develops with practice. Since the orbits are frequently not known very precisely, the comet is likely to be some distance (often many arcminutes) away from its predicted position. Therefore, in moderate to large telescopes, it is often necessary to sweep the telescope and search for the comet. Smaller telescopes usually have sufficient field of view even through a photometer that sweeping is not necessary, but one still has to search within the field.

A much more difficult problem arises because of the inability to directly relate predicted magnitudes of comets to what is seen in the telescope. There are two sources for this problem. The first is simply lack of knowledge about the comet; often we do not know how the magnitude will change in the future. This is not usually a problem for short period comets that have been observed at recent apparitions. It is a problem, however, for newly discovered comets (as was dramatically pointed out by the grossly incorrect predictions for Comet Kohoutek in 1973) and for periodic comets which have longer periods or which have not been observed at recent apparitions. More importantly, even for well-observed, short-period comets, the total magnitude, m_1, often bears little relationship to what is seen at the eyepiece, particularly in larger telescopes. This is due to the fact that comets exhibit a wide range in their degree of central condensation. It generally turns out that very dusty comets exhibit a sharply peaked distribution. In this case, the brightness seen in the eyepiece is reasonably well correlated with m_1 and additionally m_1 is not much brighter than m_2. Comets without much dust, however, exhibit a much flatter brightness distribution (cf. the discussion on scales in Section III). For these comets, the visual appearance is often much fainter than implied by m_1 because the eye does not do well at adding up the light over a large solid angle. In these comets, also, the difference between m_1 and m_2 is large (perhaps 5 magnitudes) such that the only safe prediction is that the visual appearance will correspond to something in the range from m_2 to m_1. This, unfortunately, leaves the photometrist rather unsure of what to look for unless he has observed this particular comet previously.

Once the comet has been found, it is a major advantage to have offset rates available for the telescope drives in both right ascension and declination. For comets near the earth, the deviations from sidereal motion may be so large that the comet will drift across a significant fraction of the aperture in less than a minute of time. Without offset tracking rates, this may require recentering after only a few tens of seconds of integration. Since the orbits of very long-period comets are oriented randomly, these drifts can occur in any direction and are not confined to motion along the ecliptic. As an extreme example, Comet IRAS-Araki-Alcoch 1983d had a motion of 2 arcsec per second, largely in declination. Lack of offset rates does not preclude observations; it just makes them much less convenient and efficient.

Sky Subtraction

There are two possible pitfalls which must be avoided in measuring the sky near the comet. The first is that the sky brightness is extremely variable.

In twilight, when cometary observations are frequently made, the overall sky brightness is changing rapidly. In addition, the natural sky glow at night, which consists primarily of emission features of O, O_2, and OH plus the Na-D lines, varies by large factors on time scales of tens of minutes to hours. It is therefore essential that sky brightness measurements be made more often than might be standard for stellar photometry. It is frequently necessary to have sky readings bracketing the cometary readings in order to interpolate the temporal changes in the sky and if the twilight is significant, non-linear interpolation using three or more determinations may be required.

A much more common pitfall in cometary photometry is to take sky readings too close to the comet. Regardless of the visual appearance of the comet, some emission features are likely to extend 10^6 km from the nucleus. For example, in a dusty comet, one might well find strong CN emission, to which the eye is not very sensitive, extending far beyond the visual image which emphasizes the dust concentrated near the nucleus. It is therefore not sufficient to offset until the visual image is seen to lie outside the diaphragm, particularly when emission bands are being measured. It is essential to offset by many arcminutes, even up to a degree in some cases. Taking sky readings too close to the comet is probably the most common pitfall in cometary observations made by professional astronomers who are often constrained, by instruments designed for observations of point sources, to measuring the sky at a fixed distance from the nucleus.

Extinction Corrections

The extinction corrections present perhaps the most difficult aspect of cometary photometry because comets--particularly newly discovered bright ones--frequently must be observed at large air masses, and because the observations cannot be repeated on a "better" night (since the comet will have changed).

Because the air masses are large, the first requisite is an accurate determination of the air mass at which the observation is made. This will often require knowledge of the local hour angle to better than one minute of time since the air mass changes rapidly with time at large hour angles. Note that a newly discovered comet, for which only an approximate orbit is known, is likely to have a true right ascension which differs by minutes of time from the ephemeris position. One should be careful, therefore, to apply cross-checks on the position of the comet to be sure that the local hour angle is correctly computed. This can, of course, be checked (and recalculated if necessary) long after the observation whenever a better set of orbital elements becomes available. Once the local hour angle is accurately known, it is also necessary to use a suitable formula for air mass. Although $x = \sec z$ is an excellent approximation for air mass, $x < 2.0$, cometary observations are frequently made at much larger air masses. Hardie (1962, p. 180) has given a useful expression: $x = \sec z - 0.0018167 (\sec z - 1) - 0.002875 (\sec z - 1)^2 - 0.008083 (\sec z)^3$. We (R. L. Millis and I) have used this expression and found it to linearly fit extinction data obtained in Flagstaff, Ariz., beyond $x = 8$ during our observations of Comet West in 1976. Figure 3-9 shows an example of those results with rms scatter of $0^m\!.024$ about the least squares line.

The large air masses involved here also suggest special techniques for determining the extinction. Since there are numerous local factors that can affect the extinction along a nearly horizontal path, an extinction star should be chosen in the same part of the sky as the comet, preferably one at nearly the same declination as the comet and either about one-half hour ahead of a morning comet or one-half hour behind an evening comet. Interspersing observations of the extinction star with those of the comet will then allow precise observations in conditions that are less than ideal. The only constraints on choosing the local extinction star are that it not be a variable and that it have enough flux at the wavelengths of interest (e.g., a blue star for measurement of OH).

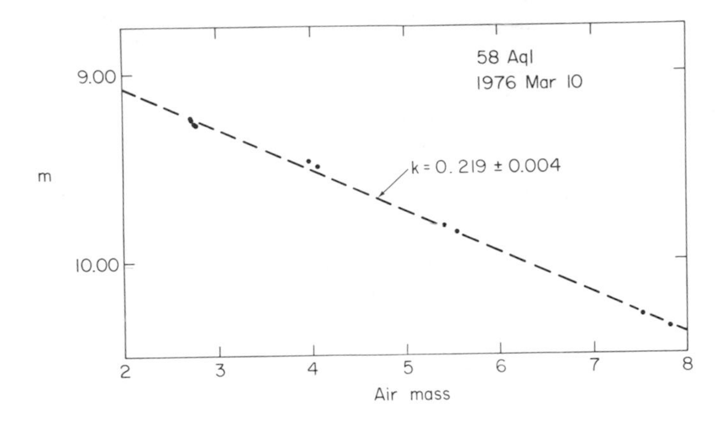

Figure 3-9. Atmospheric extinction at large air masses. Observations of the Star 58 Aql, used as an extinction star near Comet West 1976VI, through a filter at 4700 Å on 1976 March 10. Airmass was calculated using Hardie's polynomial given in text. Data taken by Millis as part of observations described by A'Hearn et al. (1977).

The extinction star can then be related to standard stars on the same night if conditions are good or on some other night when conditions are better and the extinction star can be observed at small air masses.

A final complication occurs for observations with filters at the OH band (3085 Å). It is well known among stellar photometrists that the extinction coefficients for broadband photometry (e.g., UBV) depend on the color of the star, i.e., on the flux distribution within the bandpass of the filter, because the monochromatic extinction varies within the bandpass. The filters used for cometary photometry are usually sufficiently narrow that the monochromatic extinction can be considered constant within the bandpass. This is not the case for filters shortward of 3200 Å. Not only is the extinction coefficient different for B-stars, G-stars, and comets, but for the stars the net extinction is not even linear with air mass. Details of the extinction corrections for this unusually difficult case have been given by A'Hearn, Millis, and Birch (1981).

Standard Stars

Cometary photometry is probably unique among the branches of photometry inasmuch as it deals with two different types of problems which ideally require completely different sets of standard stars. It is like nebular photometry in measuring the absolute flux of emission features, a measurement which requires standard stars for absolute flux calibration, but it is also like asteroidal photometry in measuring color differences with respect to the Sun, a measurement which can in principle be done with these same standard stars but which is commonly and much better done with standards that mimic the Sun. We will address these two types of standard stars separately.

Standard stars similar to the Sun are needed particularly for comets because the color differences between comet and Sun are not large. Since cometary photometry is done with many different types of filters, the standards must mimic the Sun in many different respects. Similarity of the B-V color is certainly not enough. Fortunately, this question has been discussed in a series of papers by Hardorp (1982 and references therein). He has identified seven stars which match the Sun both in their overall energy distribution from 3308 Å to 8390 Å and also in the strength of discrete absorption features at 3740 Å and 3850 Å. Note that the strength of the feature at 3850 Å is particularly important for cometary observations near the CN band (the solar feature is also due largely to CN). There are also questions about which of the several determinations of solar energy distribution should be used, but Hardorp has addressed these questions in some detail.

The stars he has selected are given in Table 3-2. Unfortunately, the stars are not well distributed around the sky so a proper photometric system should have secondary

standards distributed around the sky which are reasonably similar to the Sun but need not be identical. Lists of such stars can be found in the same papers by Hardorp and a particular set of 5 secondary standards used for asteroidal photometry has been given by Tedesco _et al_. (1982). The standards in Table 3-2, as observed with our comet filters, all have identical colors within $\pm 0^{m}_{.}01$.

TABLE 3-2

Solar Analogs

HD	Other Ident	α (1950)	Δ	V	B-V	Spectrum+
28099	Hyades - vB 64	4:23:47.7	+16°38'07"	8.12	+0.66	G6V
29461	Hyades - vB 106	4:36:07.6	+14°00'29"	7.96	+0.66	G5V
30246	Hyades - vB 142	4:43:38.9	+15°22'59"	8.33	+0.67	G5V
44594	HR (YBS) 2290	6:18:47.1	-48°42'50"	6.60	+0.66	G2V
105590*	BD -11° 3246	12:06:53.2	-11°34'36"	6.56	+0.66	G2V
186427**	16 Cyg B	19:40:32.0	+50°24'03"	6.20	+0.66	G5V
191854AB++	BD + 43° 3513	20:08:33.7	+43°47'44"	7.42	+0.66	G5V

+Spectral types are from various sources and apparently indicate differences in the classifiers rather than in the stars.

*Brightest member of a triple system; $8^{m}.9$ and $9^{m}.1$ companions at ($1^{s}.6$ E, 4" N) and ($0^{s}.5$ W, 22' S).

**Fainter member of a double system; 6^{m} companion at (3^{s} W, 28" N).

++Unresolved double.

The establishment of emission line fluxes requires very different standard stars. Traditionally, Vega (α Lyrae) has been used as a fundamental standard star, particularly for spectrophotometry; numerous lists of secondary standards exist, most of which have been derived ultimately from Vega. Unfortunately, many of these standard stars have spectral types similar to that of Vega, viz late B or early A. While these stars are quite suitable as spectrophotometric standards, the presence in these stars of a very strong Balmer jump, which happens to coincide with the strongest CN band in comets, makes them unsuitable standards for filter photometry in the spectral region from 3500 Å to 4000 Å. The reason for this is that the stellar spectrum is very steep in this region; thus the flux through a typical filter for cometary photometry depends sensitively on the exact position of the filter bandpass. In fact, since the filter bandpass changes systematically with temperature, these stars will systematically seem brighter on summer nights than on winter nights. Accurate calibration of fluxes, therefore, requires more accurate knowledge of the filter's characteristics than is usually available.

The obvious alternative is to use standard stars that have a weak Balmer jump and, more generally to avoid analogous problems at other wavelengths, have no strong spectral features. The stars with the most featureless spectra are of types late O and early B. For our own work, we have picked stars in the range from O9 to B4. These stars also have the advantage of having significant flux in the ultraviolet making them suitable for use in calibrating measurements of the OH feature at 3085 Å.

For our own work we have picked a set of moderately bright stars which are uniformly distributed around the celestial equator plus one star that is very far north. This set is given in Table 3-3. The flux distributions of four of these stars (μ Tau, η Hya, η UMa, and 96 Her) have been measured by several investigators and been published in the catalogue by Breger (1976). The remaining stars in the list can be established as secondary standards by photometry through the filters of interest. No attempt has been made to get identical stars for this list so there is a range of intrinsic energy distributions as well as a range of amounts of interstellar reddening. Observers with large telescopes may find some of these stars too bright, but for the majority of observers the range of brightness in this list is ideal, except for the case of η UMa which may be too bright even with moderate-sized telescopes.

As in any photometric program, it is valuable to observe several standards each night in order to catch errors, instrumental changes, transparency anomalies, etc.

TABLE 3-3

Flux Standards

HD	Other Ident	α (1950)	Δ	V	B-V	Spectrum
3379	53 Psc	00:34:10.8	+14°57'24"	5.88	-0.15	B2.5IV
26912*	μ Tau	04:12:49.0	+ 8°45'07"	4.27	-0.07	B3IV
52266	BD -5° 1912	06:57:53.9	- 5°45'21"	7.23	-0.01	09V
74280*	η Hya	08:40:36.7	+ 3°34'46"	4.29	-0.20	B4V
89688	(RS) 23 Sex**	10:18:27.1	+ 2°32'31"	6.68	-0.09	B2.5IV
120086	BD -1° 2858	13:44:44.2	- 2°11'40"	7.89	-0.18	B3III
120315*	η UMa	13:45:34.3	+49°33'44"	1.86	-0.19	B3V
149363	BD -5° 4318	16:31:47.9	- 6°01'59"	7.80	+0.01	B0.5III
164852*	96 Her	18:00:14.7	+20°49'56"	5.27	-0.09	B3V
191263	BD +10° 4189	20:06:15.1	+10°34'44"	6.33	-0.14	B3IV
219188	BD +4° 4985	23:11:28.0	+ 4°43'29"	6.9	-	B0.5III

*Flux distribution published by Breger (1976).

**Although it has a variable star designation, several more recent investigations find $\Delta m < 0^{m}.01$.

Summary Guidelines

- -Expect the comet brightness in the eyepiece to be different than predicted.

- -If possible, use offset tracking at the comet's rate.

- -Measure sky brightness at large distances from the comet.

- -Measure extinction in that part of the sky where the comet is located.

- -Use solar standards or very hot standards (or both) depending on the program.

V. DATA REDUCTION

In this section, we will address only a few points in the reduction of raw data. We will not discuss physical interpretation. The corrections for atmospheric extinction were discussed in Section IV together with the discussion on measuring extinction and will not be discussed further here.

Establishing Standard Magnitudes

The instrumental magnitudes (-2.5 log of output after correction for amplifier gain and extinction) can be converted to a standard system by addition of an arbitrary zero point shift (which can be different for each filter), as in any photometric system. The standard system, however, must first be established for each program of photometry. As examples, A'Hearn and Millis have established a standard system for their work (cf. Section II) and B. Zellner is in the process of establishing a standard system for use with the IAU-sponsored standard filters. Any other set of filters to be used on numerous comets would need a similar standardization. The process is sometimes complicated because the instrumental response rarely remains constant from one night to another. Fortunately, the process is simplified by the fact that it is only differences between the comets and the standards that are finally of interest. This allows for numerous variations in approach and we describe one approach here.

Observations of as many standards as practical are made on several nights. All of the data are reduced to instrumental magnitudes outside the atmosphere. Choose any "good" night (many standards observed, good photometric conditions, etc.) as a starting point. For each filter separately and for each remaining night separately, find the average (over all standard stars) magnitude difference between the instrumental magnitude on the night chosen as the starting point and the instrumental magnitude on the night in question. Add these average magnitude differences to the instrumental magnitudes of all stars in each filter and on each night. Then, for each filter and each standard star, average all the nights together.

At this point it is appropriate to introduce the arbitrary zero points. There are numerous ways of doing this including (1) choosing the magnitudes (in all filters) of one star to be $0^m.00$ or any other convenient number, (2) choosing the average magnitudes of some group of stars (e.g., the group with flux distributions published) to be $0^m.00$ or any convenient number, (3) choosing the zero point such that magnitudes correspond approximately to UBV magnitudes, and (4) choosing the zero point such that magnitudes correspond approximately to V magnitudes combined with the monochromatic magnitudes tabulated in flux distributions. The choice among these options is largely a matter of aesthetics although occasionally there are also practical considerations if one ultimately wants to convert magnitudes to fluxes. However, once it is chosen, addition of the zero point effectively establishes the standard magnitude system.

To completely specify the system of standards, observations must be spread over a year, dropping the westernmost stars and adding additional stars further east as the seasons progress. Ultimately, some of the stars observed initially in the western sky in the evening will again be observable in the eastern sky in the morning. If these stars are treated as newly observed stars, the resultant magnitudes can be compared with the original standard magnitudes to check for proper closure of the system. During the course of the year, data will continue to accumulate on some of the original standards thereby improving the precision of the results. The standard system can be permitted to evolve in this initial period by including the new data (as a weighted average) and adjusting the arbitrary zero point to maintain the initially desired condition. After that initial year or so, the system should be considered as defined by the observations of standard stars up to that point. All subsequent observations of comets are reduced to the standard system by determining the additive constants needed to bring the standard stars' magnitudes into agreement with the standard system. Because the filters will usually have narrow bandpasses, the transformation to the standard system should not require a color dependence as is often required in broadband photometry.

Flux in an Emission Feature

For the purpose of this discussion, let us initially assume that the filters for the continuum sample purely the continuum, i.e., that they are not contaminated by the tails of any emission bands. Suppose we have two continuum bandpasses, c1 and c2, and one bandpass centered on an emission feature, E, and we want to determine the flux in the emission band. The first step is to correct for the continuum flux under the emission band. This is most easily done by using the solar analogs.

Let the magnitude at the emission band be represented by

$$m(\lambda_E) = A + \frac{\lambda_E - \lambda_{c1}}{\lambda_{c2} - \lambda_{c1}} m(\lambda_{c2}) + \frac{\lambda_{c2} - \lambda_E}{\lambda_{c2} - \lambda_{c1}} m(\lambda_{c1}),$$

where λ_E, λ_{c1}, and λ_{c2} are the effective wavelengths of the three filters. From the observational data find A for each of the solar analogs (these values should all turn out to be the same within $\pm 0^m.01$) and determine the average value, $\bar{A}$. Once the standard system has been established, this value should never change. We can now use $\bar{A}$ to determine the cometary magnitude, $m_{c☄}(\lambda_E)$ which would have been observed in the absence of any emission band:

$$m_{c☄}(\lambda_E) \equiv \bar{A} + \frac{\lambda_E - \lambda_{c1}}{\lambda_{c2} - \lambda_{c1}} m_{c☄}(\lambda_{c2}) + \frac{\lambda_{c2} - \lambda_E}{\lambda_{c2} - \lambda_{c1}} m_{c☄}(\lambda_{c1}),$$

The magnitude of the comet due solely to the emission feature alone is then given by

$$m_{E☄}(\lambda_E) = -2.5 \log [10^{-0.4m_{☄}(\lambda_E)} - 10^{-0.4m_{c☄}(\lambda_E)}]$$

This method removes the underlying continuum including all wavelength-dependent structure if the solar analogs are truly like the Sun and if the log of the albedo of the cometary material varies linearly with wavelength over the interval including λ_E, λ_{c1}, and λ_{c2}. The latter condition is usually satisfied because the albedo itself does not vary strongly with wavelength and, as far as is known from previous comets, any discrete features are weak.

The next step is to convert the emission band magnitude, $m_{E☄}(\lambda_E)$, to the total flux in the emission band. The hot standard stars with measured flux distributions yield directly the relationship between effective continuum flux and magnitudes in the standard system, i.e., the value of B in

$$m_*(\lambda) = B(\lambda) - 2.5 \log F_{\lambda *}(\lambda).$$

The average value of B, $\overline{B}$, is used to determine the effective continuum flux equivalent to the observed emission band flux from the comet:

$$F_{eff☄}(\lambda_E) = 10^{0.4[B(\lambda_e)-m_{E☄}(\lambda_E)]}.$$

The final step requires a knowledge both of the transmission curve, $T(\lambda)$, of the filter at the emission band and of the profile $I_\lambda(\lambda)$ of the emission feature being measured. The latter is usually available from published spectrophotometry (by assuming that the profile does not change appreciably from one comet to another) while the former must be measured in the laboratory. If both these quantities are available, then the net flux in the emission feature, F_E, is given by

$$F_E = F_{eff☄}(\lambda_E) \frac{\int I_\lambda(\lambda)d\lambda \int T(\lambda)d\lambda}{\int I_\lambda(\lambda)\, T(\lambda)\, d\lambda}$$

All of the integrals can be evaluated numerically and the ratio of the integrals is then a constant of the photometric system unless either the filter ($T(\lambda)$) or the emission feature ($I(\lambda)$) changes. In evaluating the integrals, note that $\int T(\lambda)d\lambda$ has the dimensions of wavelength which must agree in units with the units of $F_{eff☄}(\lambda_E)$ which has the dimensions of flux per wavelength. The other integrals are dimensionless.

This entire procedure, however, will fail if the continuum filters are contaminated by portions of emission features. If the contamination is due to an emission feature whose flux has been measured, the correction is tedious but not difficult. As an example, filters for the blue or green continuum are almost always contaminated by the tails of C_2 bands, but one is usually also measuring a C_2 band in the photometric system. If the flux in an emission feature, F_E, has been determined, the amount of flux entering as a contaminant in an adjacent filter for the continuum is

$$F_{\lambda\ contam} = F_E \frac{\int I_\lambda(\lambda)\, t(\lambda)\, d\lambda}{\int I_\lambda(\lambda)\, d\lambda \int t(\lambda)\, d\lambda},$$

where $t(\lambda)$ represents the transmission curve of the continuum filter. This can then be used to derive a small correction to the originally observed magnitude for the continuum such that

$$m_{c☄}(\lambda_c) = -2.5 \log \left\{ 10^{0.4[m_{☄}(\lambda) - \overline{B}(\lambda_c)]} - F_{\lambda,\mathrm{contam}} \right\}$$

At this point, one can iterate through the procedure, re-estimating $m_{E☄}(\lambda_E)$ from $m_{☄}(\lambda)$, $m_{c☄}(\lambda_{c1})$, and $m_{c☄}(\lambda_{c2})$. If the filters have been well designed, this procedure can be considered entirely converged after one iteration. The method works, of course, only if the contamination of the continuum is due to a feature that has itself been measured (or which can be inferred directly from a measured feature such as the $\Delta v = +1$ sequence of C_2 being inferred from the $\Delta v = 0$ sequence). It often happens in comets with very little dust that most of the emission in the continuum bandpass is due to numerous weak emission features that are not normally measured. In this case, there is no way to correct for them and no meaningful continuum magnitudes can be derived. The corrections for the continuum underlying an emission feature also cannot be done correctly but the problem occurs only when the continuum corrections are very small so that this does not represent a significant limitation.

Continuum Colors

If the continuum magnitudes $m_{c☄}(\lambda_c)$ have been derived as described in the preceding section or if the emission band corrections are negligible so that $m_{c☄}(\lambda_c) = m_{☄}(\lambda_c)$, then the colors of the cometary materials are most easily reported as color excesses:

$$E(\lambda_{c1}, \lambda_{c2}) = [m_{c☄}(\lambda_{c1}) - m_{c☄}(\lambda_{c2})] - [m_{\odot}(\lambda_{c1}) - m_{\odot}(\lambda_{c2})]$$

where the solar color is taken to be the average color of all the solar analogs:

$$m_{\odot}(\lambda_{c1}) - m_{\odot}(\lambda_{c2}) = \overline{m_{G*}(\lambda_{c1}) - m_{G*}(\lambda_{C2})}.$$

As noted previously, all the solar analogs should have nearly identical colors.

It is possible to convert the color excess into the average slope (vs. wavelength) of the albedo of cometary material but unless one is doing a comparison with actual reflectivity curves, this last step is unnecessary.

Diaphragm Corrections

As discussed in Section III regarding choice of diaphragms, the only realistic way to compare comets is to compare them as measured within an aperture of fixed linear size (e.g., in km) when projected at the comet. In the absence of an iris that can be set to subtend a fixed linear dimension at the comet, these corrections must be made numerically. Making them, however, requires assumptions about the spatial distribution of brightness.

It is commonly assumed that the continuum flux is directly proportional to the diameter of the diaphragm, an assumption that has been shown to be very accurate in many instances but very wrong in a few instances. The distribution of the light of an emission band is much more complicated and is beyond the scope of this chapter. The reader wishing to pursue this topic further can refer to any of a large number of articles describing physical models of the cometary coma. A commonly used model is that of Haser (1957). Use of this model to correct observations to an infinitely large diaphragm has been discussed by A'Hearn (1982), by Meisel and Morris (1982), and in numerous references therein. The interested reader can obtain many of the relevant formulae from these references. As noted in Section II, much of that work involves extrapolation to a diaphragm of infinite size followed by division by a lifetime to determine production rates. At this point, however, one is doing physical modeling rather than data reduction and we will not address these questions.

VI. SUMMARY AND OUTLOOK

The techniques of cometary photometry can be readily implemented at almost any observatory since the equipment is very standard. Only minor modifications to photometers would usually be required. The major step would be acquisition of appropriate filters. The smaller observatories have, at least for sufficiently bright comets, a major advantage because it is only at such an observatory that one can usually schedule programs that require repeated observations on many nights.

The coming of Comet P/Halley (1982i), with perihelion on 9 February 1986, will provide one of the few chances to plan far ahead for an observing program on a bright comet. Nevertheless, these programs can be applied to any comet and the smaller observatories are often in a much better position than are large observatories to respond quickly to newly discovered comets. It is in this area, photometry of newly discovered, brighter comets, that smaller observatories should play a key role.

REFERENCES

A'Hearn, Michael F. (1975). "Spectrophotometry of Comet Kohoutek 1973f - 1973XII," A.J. 80, 861875.

________ (1981). "Chemical Abundances in Comets" in Comets and the Origin of Life, ed. C. Ponnanperuma (D. Reidel), pp. 53-61.

________ (1982). "Spectrophotometry of Comets at Optical Wavelengths" in Comets, ed. L. L. Wilkening (University of Arizona Press), pp. 433-460.

________, and Millis, Robert L. (1980). "Abundance Correlations Among Comets," A.J. 85, 1528.

________, and Millis, Robert L. (1983). "OH Abundances in Comets," in preparation.

________; Millis, Robert L.; and Birch, Peter V. (1981). "Comet Bradfield 1979X: The Gassiest Comet?," A.J. 86, 1559-1566.

________; Millis, Robert L.; Schleicher, D. G.; and Feldman, P. D. (1983). "Vaporization in Comet Bowell (1980b)," in preparation.

________; Thurber, Clifford H.; and Millis, Robert L. (1979). "Evaporation of Ices from Comet West," A.J. 82, 518-524.

Breger, Michel (1976). "Catalog of Spectrophotometric Scans of Stars," Ap.J. Supp. 32, 7-87.

Cochran, A. L.; Cochran, W. D.; and Barker, E. S. (1982). "Spectrophotometry of Comet Schwassman-Wachmann 1. II. Its Color and CO^+ Emission," Ap.J. 254, 816-822.

Fay, Theodore, Jr., and Wisniewski, Wieslaw (1978). "The Light Curve of the Nucleus of Comet d'Arrest," Icarus 34, 1-9.

Hall, D. S., and Genet, R. M. (1981). "Photoelectric Photometry of Variable Stars," (IAPPP Communications, Fairborn, Ohio).

Hardie, Robert H. (1962). "Photoelectric Reductions" in Astronomical Techniques, ed. W. A. Hiltner (University of Chicago Press), pp. 178-208.

Hardorp, J. (1982). "The Sun Among the Stars, V. A Second Search for Solar Spectral Analogs. The Hyades Distance," Astron. Astrophys. 105, 120-132.

Haser, L. (1957). Bull. Acad. Roy. Sci. Liege 43, 740.

Lyttleton, R. A. (1951). "On the Structure of Comets and the Formation of Tails," M.N.R.A.S. 111, 268-277.

Meisel, David D., and Morris, Charles S. (1982). "Comet Head Photometry: Past, Present, and Future," in Comets, ed. L. L. Wilkening (University of Arizona Press), pp. 413-432.

Millis, R. L.; Ahearn, M. F.; and Thompson, D. T. (1982). "Narrowband Photometry of Comet P/Stephan-Oterma and the Backscattering Properties of Cometary Grains," A.J. 87, 1310-1317.

Roach, F. E., and Gordon, Janet L. (1973). The Light of the Night Sky (D. Reidel).

Schmidt, M., and van Woerden, H. (1957). "The Intensity Distribution of Molecular Bands in the Coma of Comet Mrhos 1955e," Mem. Soc. Roy., Sci. Liege, 4e Ser., 18, 102-111.

Sekanina, Zdenek (1976). "A Continuing Controversy: Has the Cometary Nucleus Been Resolved?" in The Study of Comets, ed. B. Donn et al. (NASA SP-393, U.S. Government Printing Office), pp. 537-585.

Tedesco, Edward F.; Tholen, David J.; and Zellner, Ben (1982). "The 8-Color Asteroid Survey: Standard Stars," A.J. 87, 1585-1592.

Thompson, D. T.; Millis, R. L.; and A'Hearn, M.F. (1982). "The Disappearance of OH from Comet Encke," B.A.A.S. 17, 754 (abstract).

Whipple, Fred L. (1978). "Cometary Brightness Variation and Nucleus Structure," Moon and Planets 18, 343-359.

Zellner, B.; Tedesco, E.; and Degewij, J. (1979). "Periodic Comet Tempel 2 (1977d)," IAU Circular 3326.

4. LUNAR PHOTOMETRY

Peter Hedervari

I. INTRODUCTION

In recent years some earthbound astronomers have expressed rather pessimistic opinions about lunar research. "What can we do with our modest telescopes and other equipments in the field of lunar research after the excellent results obtained by the manned and unmanned lunar expeditions?"

I do not share this opinion. There still are interesting lunar problems which are unsolved as yet and, consequently, there are possibilities for lunar observers. For instance, scientifically valuable research can be carried out by observing lunar transient events such as brightenings or fadings of certain areas, discolorations of crater floors, etc. In addition to the lunar transient phenomena, the detection after real physical changes on the Moon's surface (landslides, for example) is also an important task. Lunar observers may wish to investigate the variation of the appearance of domes (by all likelihood they are true basaltic shield volcanoes) as the function of the degree of their illumination by the Sun. Finally, photometric measurements of selected lunar areas may be an extraordinarily fruitful scientific project. The idea of such research has been suggested by the author to the members of the Lunar Section of Association of Lunar and Planetary Observers (ALPO) (Hedervari, 1982a).

For those who are interested particularly about lunar photometry, the following publications are recommended as further readings: Struve (1960), Minnaert (1961), Barbier (1961), Fielder (1961), Baldwin (1963), Fielder (1965), Westfall (1972, 1975, 1979, 1980). And, as a particularly important source, the paper by van Diggelen (1959) is also suggested.

II. DEFINITIONS

Lunation Curve. A plot of intensity versus phase angle shown in a Descartesian coordinate system.

Phase Angle (denoted by "g"). This is the angle at the Moon's center, between the direction of the center of the Earth and Sun. The value of g varies in the following manner:

	First quarter	Full Moon	Third quarter	New Moon
g =	- 90°	0°	+ 90°	±180°

The Cardinal Points. When you are looking at the Moon through a terrestrial-type telescope (or with your naked eye), the north is at the top, and the east is on the right in the direction of Mare Crisium basin. The orientations to the east and west is based on a new definition, accepted by the International Astronomical Union in 1961. It is important to note, that prior to the birth of this definition, the east and west were just the opposite. Accordingly, in terms of this classical view, the west was on the right in the direction of Mare Crisium. You will find east and west in this meaning in papers and books that have been published before 1961.

Selenographic Coordinates. The position of points on the lunar surface is determined by the selenographic longitude and latitude, denoted by λ and β, respectively. The longitudes are measured in the plane of the Moon's equator, the axis of reference being the radius of the celestial body that passes through the visible center of the lunar disc. The positive direction is towards the east (IAU-meaning), i.e., towards Mare Crisium. The latitudes are measured from the equator towards the poles. It is positive towards the north (into the hemisphere containing crater Plato). On the visible hemisphere of the Moon the longitudes range from -90° through 0° to +90° where -90° means the western limb (IAU-meaning).

Albedo. It is the ratio of the total amount of light reflected from a body in all directions to the total incident light. The albedo of the total visible hemisphere of the Moon (Baldwin 1963) is 0.073, while the albedo of the darkest areas on the Moon in case of total illumination is 0.040. It follows from the definition that the instantaneous albedo of those parts of the lunar surface which are in perfect shadow is practically 0.0. Table 4-1 gives the albedo of larger craters, while Table 4-2 gives the almedo of basins.

Radiance (formerly called "brightness"). It is the radiation of a unit projected surface in the direction of the Earth, measured per unit solid angle. It is the direct result of the observation and it depends on the phase angle g. If the radiance is plotted as a function of the phase angle we find a lunation curve for the area considered (this definition is due to van Diggelen, op. cit., p. 13). The symbol of radiance is R. The relationship between magnitude and radiance is as follows:

$$m = - 2.512 \log R.$$

The radiance informs us about the relative brightness of the measured area. The basic value to which all the measured ones should be compared is R = 1.0. The corresponding magnitude is

TABLE 4-1

ALBEDO OF LARGER CRATERS

	van Diggelen	Sytinskaya	Class +/
Albategnius	0.111	0.112	5
Alphonsus	0.107		5
Archimedes	0.081	0.088	5
Aristarchus	0.152	0.176	1
Aristoteles	0.105	0.110	1
Aristillus	0.080		1
Arzachel	0.112		3
Autolycus	0.082		1
Billy	0.063		5
Bonpland	0.087		?
Bullialdus	0.114		1
Campanus	0.089		5
Cassini	0.110		5
Catherina	0.115		4
Clavius	0.137		2
Cleomedes	0.090		5
Copernicus	0.114	0.120	1
Cyrillus	0.110		3
Firmcus	0.069		5
Fracastorius	0.102		5
Gassendi	0.091		5
Grimaldi	0.063	0.062	5
Hevelius	0.108		5
Hipparchus	0.109		5
Julius Caesar	0.073		?
Kepler	0.102	0.100	1
Lansberg	0.115		1
Langrenus	0.110	0.144	1
LeMonnier	0.062		5
Lubiniezky	0.102		?
Lyell	0.067		?
Macrobius	0.097		1
Manilius	0.081	0.122	1
Maraldi	0.066		?
Marius	0.059		5
Menelaus	0.085	0.158	1
Mercator	0.095		5
Petavius	0.114		5
Pitatus	0.068		?
Plato	0.068	0.068	5
Posidonius	0.077		5
Proclus	0.142		1
Ptolemaeus	0.095	0.108	5
Riccioli	0.071	0.060	3
Schickard	0.087	0.078 and 0.099	5
Theophilus	0.108		1
Tycho	0.131	0.154	1
Vendelinus	0.105		5
Tycho, rays/bright streaks, running away from the crater, like spokes/		0.163	

TABLE 4-2

ALBEDO OF BASINS (AVERAGE VALUES)

Mare Crisium	0.062
Mare Fecunditatis	0.069
Oceanus Procellarum	0.051 - 0.070
Sinus Iridium	0.065
Mare Tranquillitatis	0.066
Mare Serenitatis	0.070
Mare Frigoris	0.089
Mare Imbrium	0.054 - 0.074
Mare Vaporum	0.060
Mare Nubium	0.062 - 0.073
Sinus Medii	0.054
Mare Nectaris	0.080 - 0.089
Palus Somnii	0.095
For comparison:	
upland area	0.100

(After Sytinskaya 1953)

$$m = - 2.512 \log 1.0 = 0.0.$$

Practically, R = 1.0 means the "brightness" of an area in total shadow, within which the surface features are almost invisible. (R = 0.0 is physically nonexistent since log 0.0 is mathematically meaningless.) The radiance of an illuminated area means the quantity of light relative to those places for which R = 1.0. For example, if R = 20, this means that the area in question is brighter by a factor of 20 than the region for which R = 1. The corresponding magnitude is

$$m = - 2.512 \log 20 = - 3.268$$

It is important to note that this does not mean that the area measured would be so brilliant than a pointlike celestial body the magnitude of which is - 3.268, since a lunar region, having a value of R = 1.0 and m = 0.0 is a dark one and it is not comparable with a star the astronomical magnitude of which is 0.0 (as it is well-known, a star of magnitude 0 is a very brilliant celestial body). The magnitude in this case is only a relative value which tells us how many times brighter a given region is than the area which has a magnitude 0 or a radiance of 1. Practically, an area having m = 0 or R = 1 does not mean an absolute black body from which no light comes to our eyes at all, but only a very dark region on the lunar surface.

The Photometric Function. This describes the photometric behavior of the lunar surface. It can be written in this form:

$$\phi(i, e, g),$$

where i is the angle of incident of the light, e the angle of observation, and g the phase angle. Concerning this photometric function the following very important facts were stated by Soviet experts Markov, Sharonov, Fedoretz, and Orlova:

1. All lunar craters have the same photometric function.

2. The photometric function of plains, or isolated mounts or else mountain chains is practically the same as that of the crater floors.

3. All objects on the lunar surface reach their maximum radiance either just at, or rather near the full moon time, quite independently of their position. Those features for which the maximum is slightly shifted (and thus reach their maximum brightness a little after full moon) will be discussed in more detail later.

4. The form of lunation curves depends on the selenographic longitude alone, and appears to be perfectly independent of the latitude.

Van Diggelen has added that according to the photometric investigations the lunar surface has the same microstructure everywhere, irrespective of the differences in the albedo values.

III. THE LUNATION CURVE

Table 4-3 gives the effectively measured radiance and the magnitude of 38 large lunar craters as a function of phase angle. The values were obtained by measurements from five photographic plates. These were taken by Minnaert in the summer of 1946 with the 40-inch refractor of the Yerkes Observatory at a wavelength of 550 nm. All the plates were obtained within one lunation. As can be seen, only five positive phase angles were considered, without negative ones. Thus the lunation curves one can obtain by applying these numerical values is incomplete. For many of the craters enumerated here, we can find either further numerical data in van Diggelen's work, or lunation curves, or both. These data and drawings are based on the observations of other authors. However, there are ten craters (their names are underlined in Table 4-3) for which further data does not appear to exist.

These ten craters should be the object of further investigations, and one of the possible programs for small observatories can be the completion of the lunation curves. This is practically possible since at least two values have already been established for each of these ten craters. Consider, for example, crater

TABLE 4-3

Radiance and Magnitude, after van Diggelen (1959)

Crater	Radiance, R phase angle, g (degrees)					Magnitude, m phase angle, g (degrees)				
	+11.4	+23.9	+77.9	+91.3	+104.7	+11.4	+23.9	+77.9	+91.3	+104.7
Albategnius	88.2	69.2	6.9			-4.87	-4.60	-2.08		
Alphonsus	85.8	63.3	13.1	6.7		-4.83	-4.51	-2.77	-2.07	
Archimedes	72.1	47.2	9.7	6.6		-4.64	-4.18	-2.45	-2.06	
Aristillus	77.2	53.5	11.2			-4.71	-4.32	-2.61		
Billy	64.4	45.6	20.7	16.4	13.3	-4.52	-4.15	-3.27	-3.04	-2.81
Bonpland	86.5	60.6	17.8	9.5	2.2	-4.84	-4.46	-3.10	-2.44	-0.85
Bullialdus	110.0	75.1	15.1	11.2	7.0	-5.10	-4.69	-2.93	-2.63	-2.12
Campanus	76.2	46.3	19.1	13.3	5.7	-4.70	-4.16	-3.18	-2.81	-1.89
Cassini	90.0	65.3	6.7			-4.88	-4.54	-2.04		
Clavius	130.0	94.7	21.9	17.2		-5.28	-4.94	-3.47	-3.09	
Cleomedes	59.4	21.3				-4.45	-3.52			
Copernicus	136.4	92.3	31.4	19.9	8.1	-5.33	-4.70	-3.72	-3.25	-2.28
Cyrillus	120.1	63.5				-5.19	-4.51			
Eratosthenes	120.1		28.8			-5.19		-3.63		
Eudoxus	92.3	60.6				-4.91	-4.46			
Gassendi	91.5	60.6	30.5	23.6	14.5	-4.90	-4.46	-3.69	-3.43	-2.91
Grimaldi	56.7	46.5	24.9	19.2	17.2	-4.39	-4.17	-3.49	-3.21	-3.09
Guericke	118.2	56.5	22.4	12.9		-5.17	-4.38	-3.35	-2.78	
Hevelius	99.3	88.0	32.7	29.9	25.4	-4.98	-4.86	-3.76	-3.69	-3.51
Lansberg	84.1	50.1	20.3	17.7	9.2	-4.81	-4.25	-3.25	-3.13	-2.42
Maginus	138.6	94.5	18.2	10.0		-5.35	-4.94	-3.13	-2.51	
Manilius	69.3	59.0	3.1			-4.60	-4.43	-1.22		
Marius	50.2	40.3	17.0	15.6	12.3	-4.25	-4.01	-3.08	-2.98	-2.73

TABLE 4-3 (continued)

Crater	Radiance, R phase angle, g (degrees)					Magnitude, m phase angle, g (degrees)				
	+11.4	+23.9	+77.9	+91.3	+104.7	+11.4	+23.9	+77.9	+91.3	+104.7
Maurolycus	105.3	63.7				-5.04	-4.51			
Mercator	89.3	53.8	21.2	14.2	5.4	-4.87	-4.33	-3.29	-2.88	-1.84
Petavius	100.9	12.3				-5.00	-2.73			
Piccolomini	128.5	58.0				-5.27	-4.41			
Pitatus	61.9	49.4	13.4	11.6		-4.48	-4.23	-2.77	-2.66	
Plato	51.3	34.0	9.1	7.0		-4.27	-3.83	-2.38	-2.11	
Plinius	111.8	51.2				-5.11	-4.27			
Posidonius	84.5	47.2				-4.81	-4.19			
Ptolemaeus	93.5	56.6	11.1	4.5		-4.92	-4.38	-2.59	-1.65	
Pytheas	108.2	70.3	14.5	9.2	7.1	-5.09	-4.62	-2.88	-2.41	-2.13
Reinhold	88.2	50.0	24.7	17.3	3.1	-4.86	-4.27	-3.46	-3.10	-1.22
Stöffler	142.9	102.3				-5.38	-5.03			
Thebit	108.9	71.5	13.4			-5.08	-4.64	-2.79		
Tycho	159.8					-5.50				
Walter	115.7	84.0	13.8			-5.15	-4.81	-2.83		

Eratosthenes. Its two values are given for g = 11.4° and for g = 77.9°. Suppose that you will have an opportunity to measure the radiance of Eratosthenes at a phase angle of 11.4° and at 77.9° again, and you receive two values: A and B. The first corresponds to 120.1 and the second to 28.8. On this basis you can compute your further results, C, D, E, etc., obtained by your own measurements made at other phase angles by comparing them with the known values of A and B. In case of craters Maginus, Walter, Pytheas, and Thebit, where three or more data are given in Table 4-3, the task is easier. You can construct a drawing that is a part of the lunation curve, and on this basis it is possible to obtain values by interpolation between phase angles 11.4° and 77.9°. In other words, you are not forced to make your own measurements exactly either when g = 11.4° or g = 77.9°, but you can choose any phase angle which is found within these two limits. Interpolation when only two values are known, as was the case of crater Eratosthenes, is not possible because two points on the graph are not sufficient to establish the form of the curve over this section. Three or more points are sufficient for this purpose.

IV. ELGER'S ALBEDO SCALE

A possible second program is the extension of the so-called albedo scale of Elger. This scale is useful for making fast estimates of the brightness of the different features on the Moon, but it is not very exact. Its extension to other features and modification to increase its accuracy can be recommended as a good project for the photometrist. The 0 value in Table 4-4 below refers to whichever point of time except full moon, since at full moon no shadows can be seen on the illuminated hemisphere of our satellite. But all the other values refer to full moon. Table 4-4 is a compilation provided by Marvin W. Huddleston, ALPO Lunar Recorder. It was created in the frame of the Selected Areas Program of the Lunar Section of ALPO. Three items are added by the present author on the basis of other sources.

It was mentioned earlier that there are a few craters that do not attain their maximum radiance exactly at full moon, but somewhat later. Aristarchus is one of these exceptional craters. Let us accept the lunation curve of Aristarchus (which was created by van Diggelen on the basis of the observations carried out by other authors) as the exact representative curve for the variation of brightness of this crater. This curve is an average one, and the scattered points that represent different, individual measurements were averaged. The points for the curve are given in Table 4-5.

TABLE 4-4

Elger's Albedo Scale

Degree	Examples
0.0	Totally black shadows
0.04	Rim of Bessel
0.08	Environs of Archimedes
1.0	Darkest parts of Grimaldi and Riccioli
1.5	Interiors of Boscovich, Billy, and Zupus
2.0	Floors of Endymion, LeMonnier, Julius Caesar, Crüger, and Fourier
2.1	Interior of Copernicus
2.5	Interiors of Azout, Vitruvius, Pitatus, Hippalus, and Marius
3.0	Interiors of Taruntius, Plinius, Theophilus, Parrot, Flamsteed, and Mercator
3.5	Interiors of Hansen, Archimedes, and Mersenius
4.0	Interiors of Manilius, Ptolemaeus, and Guericke
4.5	Surface around Aristillus and Sinus Medii
5.0	Walls of Arago, Lansberg, Bullialdus, surfaces surrounding Kepler and Aristarchus
5.5	Walls of Picard and Timocharis; rays of Copernicus
6.0	Walls of Macrobius, Kant, Bessel, Mosting and Flamsteed
6.5	Walls of Langrenus, Theatetus and LaHire
7.0	Theon, Aridaeus, Bode B, Wichmann, and Kepler
7.5	Ukert, Hortensius, and Euclides
8.0	Walls of Godin, Bode, and Copernicus
8.5	Walls of Proclus, Bode A, and Hipparchus C
9.0	Censorinus, Dionysius, Mösting A, Mersenius B and C
9.5	Interior of Aristarchus and LaPyrouse
10.0	Central peak of Aristarchus

TABLE 4-5

RADIANCE OF CRATER ARISTARCHUS
AS A FUNCTION OF PHASE ANGLE

g, degrees	R	days before (-) or after (+) full moon	
-40	6	-3.28	
-30	28	-2.46	
-20	54	-1.64	
-10	90	-0.82	
0	144	0	(full moon)
+10	172	+0.82	(maximum)
+20	136	+1.64	
+30	108	+2.46	
+40	88	+3.28	
+50	70	+4.10	
+60	56	+4.92	
+70	42	+5.74	
+80	32	+6.56	
+90	24	+7.38	(last quarter)
+100	18	+8.20	
+110	12	+9.02	
+120	8	+9.84	
+130	6	+10.66	

Between two phase angles, given in Table 4-5,--say between -20° and -10°, for example,--the variation of brightness can be regarded to be approximately linear. Therefore one can state the R value for whichever point of time between -3.28 and +10.66 days by interpolation.

As can be seen from Table 4-5, at full moon the radiance of Aristarchus is 144. On Elger's map this corresponds to 9.5. However, this is not the maximum, as it is attained at g = +10° when R = 172. This value would correspond to 11.3 on Elger's scale, and the central peak of the crater is still a little brighter. Being Aristarchus, the most brilliant area on the visible hemisphere of the Moon, this can be considered as a basis (a reference point) in case of practical lunar photometry.

V. MAGNIFICATION

In lunar photometry the best magnification depends on the task. When whole craters are measured, a magnification of about 80 is recommended. But, when the aim of the observation is the measurements of fine details within a crater, much stronger magnification is needed. You can enlarge the magnification of the telescope-photometer system by using a Barlow lens in front of the photometer. For the very fine details of the lunar surface, such as individual peaks in Mare Imbrium, or small, dark areas within crater Alphonsus, central peaks of certain craters, selected points of one of the bright rays of Tycho or Copernicus, etc., high magnification should be used, at least 400 or still greater.

I should mention that my observatory possesses a Model SSP Solid-State Stellar Photometer, made by Optec Inc. My own experiences show that with the Model SSP photometer the ring appears to be almost exactly identical in diameter with crater Copernicus at a magnification of 80. The true diameter of this crater is 90.7 km. Aristarchus has a diameter of some 39.2 km. With the same magnification this crater occupies a little less than half of the ring. Thus, a certain part of the lunar surface outside the wall of Aristarchus can also be seen within the ring. However, making measurements with the same telescope and applying always the same magnification, the whole area around Aristarchus which is still within the ring can be regarded as the reference point. If you wish to carry out another measurement for which you need a higher magnification, you must remeasure Aristarchus again with this new magnification. Photometers with easily changed diaphragm sizes would, of course, provide additional flexibility in lunar photometry.

VI. REFERENCE POINTS

Lunar photometry requires reference points, the brightness of which is exactly known at the time of your measurements. In addition to this you must determine the zero point. Let us deal with this latter problem first.

It may appear to be logical to measure an area which lies in shadow, but at full moon no shadow is to be seen on the Moon! You need a zero point on every evening you carry out lunar observations. To measure a small area on the sky, where no stars and other objects are present is not recommended because the light of the Moon represents a disturbing effect, particularly around full moon. The method that can be suggested for obtaining the true zero value is to cap the telescope, flip up the mirror and put the pointer to the zero value of the scale. After this, remove the cap from the telescope and begin the measurement of the reference point. It is advisable to point your telescope to crater Aristarchus, since from g = -40 to g = +130 (-3.28 days to +10.66 days relative to the full moon's occurrence) Aristarchus can be seen and at full moon it is the most brilliant region on the lunar surface.

It may happen, however, that you wish to make observations when Aristarchus is invisible, that is prior to g = -40 or after g = +130. Therefore, you must seek some other reference points as well. In Table 4-6 the radiance of some other craters are given. A few of them can be seen before the appearance of Aristarchus, and they can be observed for some days after the appearance of Aristarchus, too. That will provide you with an opportunity to measure them and compare their radiance with the measured value of Aristarchus.

Table 4-7 informs you how many days have passed since new moon until the occurrence of a certain phase angle. This table can be used as an auxiliary one for Table 4-6.

VII. CORRECTIONS

The high linearity of the photometer permits one to measure the brightness of any point on the lunar surface relative to the radiance of crater Aristarchus or another reference point. Various corrections need to be made to these measured values before they are useful.

Atmospheric extinction. The visibility of celestial objects, and thus the brightness of the lunar features as well, depends on the altitude of the celestial body in question above the horizon, the clearness of the air, etc. This is a well-known procedure and this correction has been treated elsewhere and will not be repeated here.

Variation of the Earth-Moon distance. Your results should be corrected to the average distance, which is 384,395 km between the centers of the earth and the moon. The correction, c, is

$$c = \frac{0.0677 \times 10^{-10}}{\frac{1}{t^2}},$$

where t is the instantaneous distance between the Earth and the Moon, expressed in kilometers. To obtaining the radiance for the average Earth-Moon distance, any R value must be multiplied by c

$$R_{corr}\ (t = av) = R \times c$$

TABLE 4-6

RADIANCE OF SELECTED CRATERS

g, degree	Cleomedes	Petavius	Langrenus	Vendelinus	Proclus	Grimaldi	Riccioli	Schickard	Aristarchus
-150	6	i	10	8	i	i	i	i	i
-140	8	10	11	10	i	i	i	i	i
-130	10	12	13	12	8	i	i	i	i
-120	12	14	16	14	11	i	i	i	i
-110	14	16	19	16	15	i	i	i	i
-100	17	20	22	18	19	i	i	i	i
-90	20	25	26	20	25	i	i	i	i
-80	24	30	31	24	32	i	i	i	i
-70	29	36	37	28	39	i	i	i	i
-60	34	44	44	33	48	i	i	i	i
-50	40	51	51	38	59	i	i	i	i
-40	46	59	58	44	72	i	i	i	6
-30	52	67	66	50	85	4	i	10	28
-20	60	78	75	59	99	12	i	29	54
-10	74	90	84	72	114	31	32	56	90
0	87	108	108	102	130+/	59+/	66	84+/	144+/
+10	72	94	74	60	114	57	60	80	172
+20	42	54	36	27	84	50	52	69	136
+30	2	17	i	10	43	44	46	55	108
+40	i	i	i	i	8	38	42	46	88
+50	i	i	i	i	i	34	37	38	70
+60	i	i	i	i	i	30	34	32	56
+70	i	i	i	i	i	28	32	27	42

TABLE 4-6 (continued)

g, degree	Cleo-medes	Peta-vius	Lang-renus	Vende-linus	Proc-lus	Gri-maldi	Ricci-oli	Schic-kard	Aris-tarchus
+80	i	i	i	i	i	25	29	22	32
+90	i	i	i	i	i	22	27	18	24
+100	i	i	i	i	i	20	25	14	18
+110	i	i	i	i	i	17	23	11	12
+120	i	i	i	i	i	15	22	8	8
+130	i	i	i	i	i	13	20	6	6
+140	i	i	i	i	i	11	18	4	i
+150	i	i	i	i	i	9	16	2	i

+/The maximum value in case of Proclus is 142 at g = +4.6; for Grimaldi, it is 62 /g = +5,5/; for Schickard the maximum is 85 /g = +5,5/; and for Aristarchus it is 172 /g = +10/. Letter i means that the crater is in shadow and therefore invisible, its radiance is 1 and its magnitude is 0. Data in this Table are rounded values, determinated from the lunation curves of van Diggelen.

Variation of the Sun-Moon distance. The Moon is illuminated by the Sun, and as the distance between these two celestial bodies is constantly changing, to obtain precise results one must correct for this phenomenon. The correction, according to which the distance between the Sun and Moon is taken into account, can be computed by using the following formula:

$$c_t = \frac{0.00004469 \times 10^{-12}}{\frac{1}{t^2}} ,$$

where t is the distance between the Sun and the Moon at the time of the observation. t is expressed in 10^6 km. Selected values are given in Table 4-8.

TABLE 4-8

CORRECTION ACCORDING TO THE DISTANCE BETWEEN THE SUN AND THE MOON

t , 10^6 km	c_t (rounded)	
152.491	1.0393	t_{max}
152.000	1.0326	
151.500	1.0257	
151.000	1.0189	
150.500	1.0122	

VIII. PROGRAMS

The following programs are my suggestions for smaller observatories:

-Completion of incomplete lunation curves of craters, mentioned and under lined in Table 4-3.

-Control and extension of Elger's albedo scale (see in Table 4-4.

-Creation of lunation curves for craters, not treated by van Diggelen.

-Controlling measurements and creation of lunation curves of those craters which have been discussed by van Diggelen.

TABLE 4- 7

DAYS FROM NEW MOON PHASE

g, degree	days	lunar phase and percent of illumination of the visible hemisphere		
-180	0	new moon	0	
-170	0.8202941		5.555...(ad	
-160	1.6405882		11.111	inf)
-150	2.4608823		16.666	
-140	3.2811764		22.222	
-130	4.1014705		27.777	
-120	4.9217646		33.333	
-110	5.7420587		38.888	
-100	6.5623528		44.444	
-90	7.3826469	first quarter	49.999	50.000
-80	8.2029410		55.555	
-70	9.0232351		61.111	
-60	9.8435292		66.666	
-50	10.6638233		72.222	
-40	11.4841174		77.777	
-30	12.3044115		83.333	
-20	13.1247056		88.888	
-10	13.9449997		94.444	
0	14.7652938	full moon	99.999	≈100.000
+10	15.5855879		94.444	
+20	16.4058820		88.888	
+30	17.2261761		83.333	
+40	18.0464702		77.777	
+50	18.8667643		72.222	
+60	19.6870584		66.666	
+70	20.5073525		61.111	
+80	21.3276466		55.555	
+90	22.1479407	last quarter	49.999	50.000
+100	22.9682348		44.444	
+110	23.7885289		38.888	
+120	24.6088230		33.333	
+130	25.4291171		27.777	
+140	26.2494112		22.222	
+150	27.0697053		16.666	
+160	27.8899994		11.111	
+170	28.7102935		5.555	
+180	29.5305876	new moon	0	

-Creation of lunation curves for other features, such as peaks in basins, domes, ridges, the straight wall, Alpine and Rheita Valleys, and the Cobra Head in Schroter Valley.

-Special investigation of twin craters (e.g., Messier and Pickering). Twin craters consist of two rings, very near to one another, and having almost the same diameter. Do they have the same radiance, the same lunation curves?

-Search for craters that have a maximum radiance after full moon.

-Is there any feature on the Moon that attains its maximum radiance prior to full moon?

-Creation of one or more sections of brilliancy through a continental area, or through a mare, or else both, including mountain ranges, e.g., Appennines--Mare Imbrium--Sinus Iridum--Jura mountains. The sections must be made by the measurement of selected points along a line, and it is advisable that the distance between the points to be measured must be the same.

-Creation of a map on the distribution of radiance within a larger region, e.g., Sinus Iridum. This can be done during the full moon phase. A comparison of the distribution of radiance with a geological map of the same area would be particularly valuable from the scientific point of view.

-Creation of the lunation curves for different basins or larger parts of certain basins and their comparison with the lunation curves of continents or larger parts of continents.

-Investigation of transient phenomena sites (e.g. Plato, Alphonsus, Aristarchus, etc.), particularly in the infrared.

-All of these measurements can be repeated in different wavelengths.

Finally, I wish to suggest consideration of the form of the lunation curves and some possible programs in this area. The lunation curve of the whole visible side of the Moon is a symmetrical curve. Craters, which are near the central meridian of the lunar disc, e.g. Albategnius, also show a symmetrical curve. But the lunation curves become more and more asymmetric when the distance from the central meridian increases. For example, Vendelinus, near the eastern edge of the disc (IAU-meaning), or Aristarchus, near the western edge, have definitely asymmetric lunation curves. The asymmetry means that the slopes of the curves are not the same; one of them is rather steep, while the other is gently sloping. This asymmetry was explained by Struve (1960, p. 72): "The onset of brightness depends on the phase angle at which the Sun's rays first illuminate the crater. As we have noted, all the curves have their peaks close to full moon. Hence it is obvious that before full moon the lunation curves must be steep for craters located near the eastern limb (this time this is according to the classical definition of the east and west, and not in accordance with the decision of IAU - note by P. H.), and gradual for those near the western limb (classical definition - note by P. H.). Another property of the curves is their great steepness near full moon. This is especially conspicuous for craters near the central meridian. The heights of the curves are not all the same, for they depend to a large extent on the albedo (reflectivity) of each crater floor."

It is very much conspicuous, as mentioned previously, that there are a few craters which attain their maximum radiance slightly after g = 0, that is after full moon. These craters are given in Table 4-9.

It has been suggested by van Diggelen (1959) and Baldwin (1963) that all craters with a shifted maximum are surrounded by a system of bright rays, composed of large and often very long white streaks, radiating in all directions, with the crater as center. Similarly it was suggested that all, or almost all craters with a displaced maximum are young ones (class 1 of Baldwin). Table 4-9 shows clearly, however, that these statements are incorrect. No relationship can be found with other properties of these craters, e.g. location, diameter, albedo, presence or absence of central peaks, number of peaks if any, presence of ray system or the shape of lunation curve. It is noteworthy, furthermore, that the case of crater Posidonius is somewhat doubtful since only one measurement suggests a shift of maximum radiance.

TABLE 4-9

CRATERS THAT ATTAIN THEIR MAXIMUM RADIANCE AFTER FULL MOON*

Crater	Diameter km (rounded)	Maximum radiance at g = ... degree	Albedo	Class	Central Peak (s: single, m: multiple)	Presence of ray system, and diameter of ray pattern, km	Lunation curve (S: symmetrical A: asymmetrical)
Archimedes	80	+ 5	0.081	5	0	-	S
Aristarchus	39	+10	0.152	1	s	+, 430	A
Aristillus	56	+ 5	0.080	1	m	+, 640	S
Arzachel	97	+ 4	0.112	3	s	-	A
Billy	46	+12	0.063	5	0	-	A
Catherina	100	+ 5	0.115	4	s	-	A
Clavius	230	+ 6	0.137	2	m	-	S
Copernicus	91	+10	0.114	1	m	+, 1200	S
Cyrillus	93	+10	0.110	3	m	-	A
Gassendi	110	+ 4	0.091	5	m	-	A
Grimaldi	203	+ 5.5	0.063	5	0	-	A
Kepler	32	+ 8	0.102	1	m	+, 640	A
Lubiniezky	44	+ 6	0.102	?	0	-	A
Macrobius	63	+ 5	0.097	1	m	-	A
Posidonius	101	+11(?)	0.077	5	s	-	A
Proclus	27	+ 4.6	0.142	1	m	+, 640	A
Ptolemaeus	149	+ 5	0.095	5	0	-	A
Schikard	214	+ 5.5	0.087	5	0	-	A
Tycho	86	+12	0.131	1	m	+, 3040	A

*From the lunation curves published by van Diggelen.

REFERENCES

Baldwin, R. B., 1949. "The Face of the Moon." The University of Chicago Press. Chicago.

________, 1963. "The Measure of the Moon." The University of Chicago Press. Chicago, London.

Barbier, D., 1961. "Photometry of Lunar Eclipses." Chapter 7 in: Planets and Satellites/The Solar System, Vol. III, ed. Kuiper, G. P. and Middlehurst, B. M. The University of Chicago Press, Chicago.

Fielder, G., 1961. "Structure of the Moon's Surface." Pergamon Press, Oxford, London, New York, Paris.

________, 1965. "Lunar Geology." Lutterworth Press, London.

Hédervári, P., 1982. "On the Albedos of Some Lunar Features." Communications of the Georgiana Observatory, No. 4.

________. "A Proposed Photometric Program for A.L.P.O. Lunar Observers." The Strolling Astronomer, F-8, 147-150.

Minnaert, M., 1961. "Photometry of the Moon." Chapter 6 in: Planets and Satellites, Vol. III. ed. Kuiper, G. P. and Middlehurst, B. M. The University of Chicago Press, Chicago.

Rükl, A., 1976. "Mars and Venus." A Concise Guide in Colour. Hamlyn, London, New York, Sydney, Toronto.

Struve, O., 1960. "Photometry of the Moon." Sky and Telescope 20, 2, 70-73.

Sytinskaya, N. N., 1953. Summary Catalogue of the Absolute Values of the Visual Reflecting Power of 104 Lunar Features, in Russian, Astr. Zhur. 30, 295.

Van Diggelen, J., 1959. Photometric Properties of Lunar Crater Floors. Recherches Astronomiques de L'Observatoire d'Utrecht, XIV, 2, 1-114.

Westfall, J. E., 1972. Photometry Methods and the Total Lunar Eclipse of January 30, 1972. The Strolling Astronomer 23, 7-8, 113-118.

________, 1975. Observing Lunar Eclipses. The Strolling Astronomer 25, 5-6, 85-88.

________, 1979. The Total Lunar Eclipse of September 6, 1979. The Strolling Astronomer 27, 11-12, 225-227.

________, 1980. Photographic Photometry of the Total Lunar Eclipse of September 6, 1979. The Strolling Astronomer 28, 5-6, 116-119.

5. SOLAR PHOTOMETRY

Gary A. Chapman

I. INTRODUCTION

Ground-based solar astronomy may be entering a renaissance because of several startling discoveries concerning the Sun. These discoveries include global oscillations, missing neutrinos from the core, variations in total irradiance, and, possibly, changes in radius. At the San Fernando Observatory, we are attempting to better understand variations in the total irradiance. The San Fernando Observatory (SFO) is located at the northwestern edge of the greater Los Angeles urban area in southern California. Figure 5-1 shows the San Fernando Observatory among the trees of the Upper Van Norman Reservoir. Built by the Aerospace Corporation in 1969, the observatory was donated to the California State University, Northridge in 1976 (see Mayfield et al. 1969 for the details of the telescope). In addition to solar photometry, staff and students of the SFO are studying solar magnetic fields, and solar flares. Solar flares, now known to produce surface nuclear reactions in some cases, are receiving renewed attention both from space and from the ground.

Photometry of the Sun can be very rewarding for the serious amateur astronomer. Studying the Sun has importance both to astronomy, because the Sun is a star, and also to the Earth because the Sun influences the Earth through particles and radiant energy. Since the Sun is a close star, astronomers have a chance to test detailed theories about stellar structure. Further, since the Sun is bright, serious scientific work can be carried out with a relatively small telescope. The amateur can make an important contribution to solar astronomy since the Sun is never the same from one day to the next and many professional astronomers have very limited telescope time.

Due to the high level of irradiance of the solar disk, a relatively modest aperture telescope, in the range of 15 - 30 cm, is adequate for certain kinds of high quality photometric research. For example, at the SFO, we have done most of our solar photometry with the 28 cm (11-inch) diameter vacuum reflector shown in Figure 5-2, and we plan to do much of our future photometry with a 20 cm (8-inch) folded refractor.

Framingham State College
Framingham, Massachusetts

Figure 5-1. Telephoto view of the San Fernando Observatory. The large structure on the left carries the 61/28 cm vacuum telescope, spectroheliograph, and coude observing room. The small dome contains a guided spar with a folded refractor. To the right is the elevated patrol telescope, which is a folded 15 cm (6-inch) refractor. Solar telescopes are often elevated to get them away from turbulence caused by solar heating of the ground.

Figure 5-2. View of the vacuum reflector in operation with the 28 cm telescope feeding the coude observing room. (The larger 61 cm telescope is covered.)

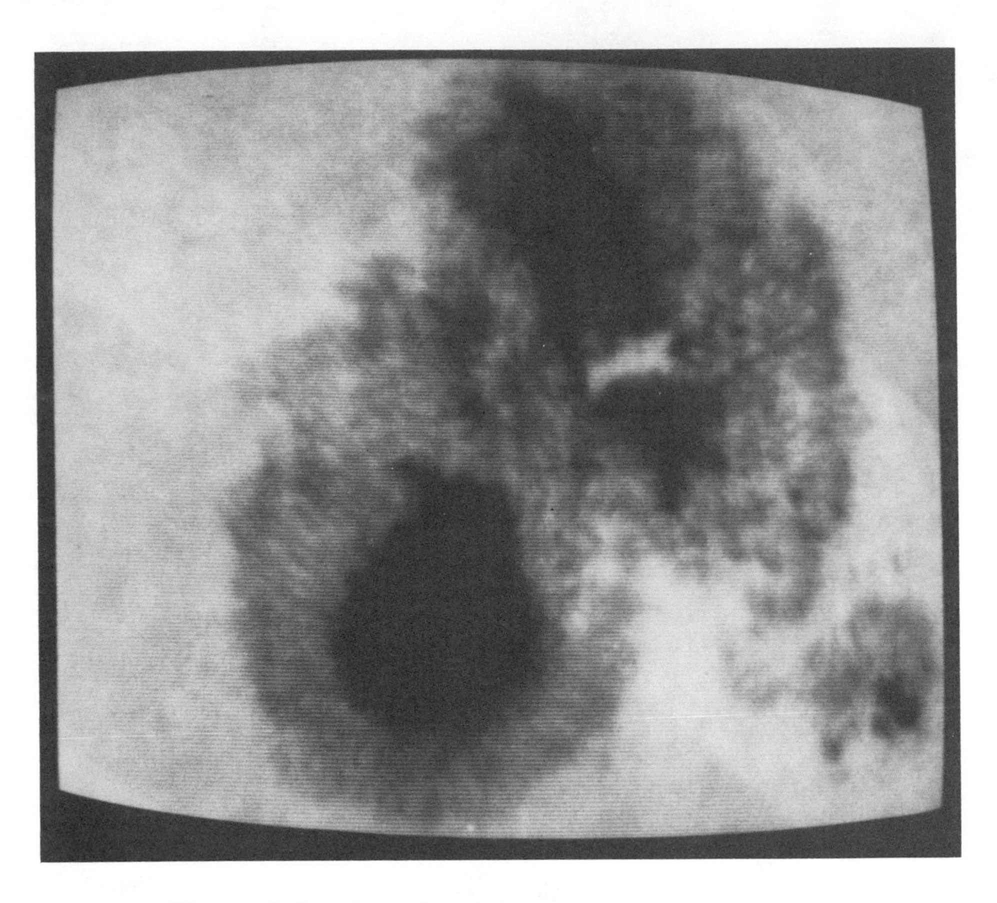

Figure 5-3. One of two large sunspots of July 24, 1981 that produced one of the largest decreases in solar irradiance seen by the ACRIM on NASA's SMM satellite. This sunspot caused a decrease of over 0.1 percent in the solar irradiance, based on photometry with the ELP.

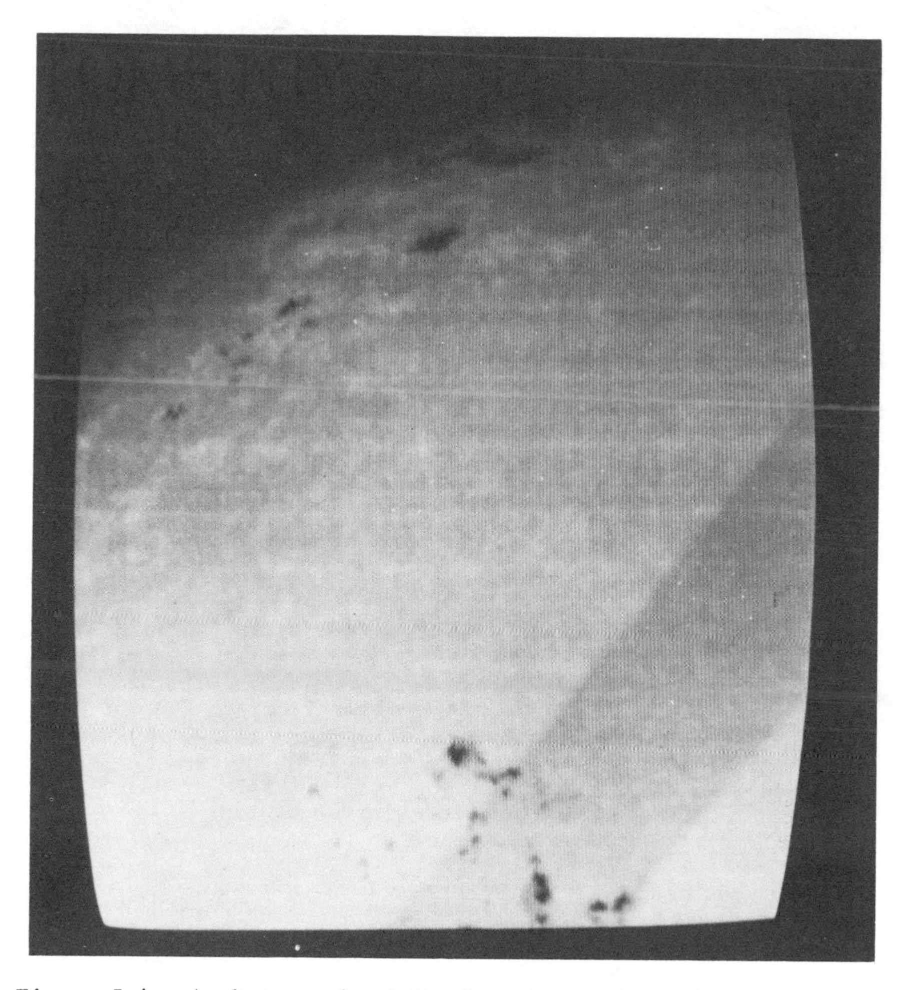

Figure 5-4. A photograph of faculae obtained in the green showing that faculae are brighter than the quiet sun but become difficult to see when not near the solar limb.

As with stellar photometry, improvements in detectors and electronics make it easier for the amateur to do serious solar photometry. In this chapter, I will describe some of the problems in solar research that may be of interest to the amateur with an interest in photometry. I will also describe what I think are some of the more interesting experiments that could be done with inexpensive equipment. Finally, I will briefly describe some of the activities at the San Fernando Observatory pertaining to photometry of the Sun.

II. ACTIVE REGION PHOTOMETRY

There are two things, quite different in nature, that can be studied by photometry. The more obvious is solar activity, in particular, sunspots. The other is the quiet Sun's limb darkening.

The magnetic field in solar active regions, so named because they are the origin of solar activity, is divided into two types of magnetic field configuration, sunspots and faculae. Sunspots are familiar to many people and are easily seen as shown in Figure 5-3 because they are dark against the photosphere due to their low temperature (about 4000K as compared to 6000K for the photosphere). Faculae, on the other hand, are less easily seen. They are visible as bright patches (as shown in Figure 5-4) but only toward the limb of the Sun. The cause of this difference between sunspots and faculae is not yet understood but has to do with the magnetic field strength. The magnetic field strength in sunspots is between 2500-3000 gauss, whereas that in faculae is only about 1500-1800 gauss. The Earth's magnetic field is about 1/3 gauss. The difficulty in seeing faculae is that they are hotter than the surrounding photosphere but only by a few hundred Kelvin at most visible wavelengths. They are also scattered out into patches rather than being concentrated as are sunspots. The mean contrast of facular regions, when taken as a whole, is about 5 percent in the midvisible, say the green part of the spectrum. This low contrast, occurring near the limb where variations in foreshortening and limb darkening are pronounced, makes it difficult to study them photometrically with a small telescope. Nonetheless, useful work on the color of faculae could be carried out with a small telescope if several color bands were measured simultaneously in order to remove the effects of seeing.

Why photometer sunspots? One reason to measure the brightness and color index of sunspots is that it provides data on the structure of sunspots because the magnetic field, velocities, and temperature are interrelated. For this work you need a lot of scale so as to obtain high resolution maps. High resolution requires a large telescope. There is another reason to study sunspots photoelectrically because sunspots cause a small but important decrease in the solar irradiance, sometimes called the solar "constant." The solar irradiance is the amount of power per square meter striking the outer atmosphere of the Earth. Much of this energy is absorbed by the ground and various parts of the Earth's atmosphere. A small change, of the order of a percent, over a long time of years, would probably have important climate effects. Although we earthlings couldn't do anything to control the Sun, we could influence our behavior and our institutions if we knew that a climate change was going to occur. This is one way in which solar astronomy differs from other areas in astronomy; the solar-terrestrial connection.

On February 14, 1980, NASA launched the Solar Maximum Mission satellite or SMM. Its purpose was to study the active Sun in a coordinated way using several different instruments. Included was a device to measure the solar irradiance at all wavelengths from X-ray into the radio. This device is called the Active Cavity Radiometer Irradiance Monitor or ACRIM for short. This device is under the control of Dr. R. C. Willson of the Jet Propulsion Laboratory. The results obtained so far show that sunspots definitely and faculae maybe control changes in the solar irradiance (Willson et al. 1981). An important question is, does solar activity cause all of the changes seen in the solar irradiance? Solar scientists are presently divided on this question. If not, if some part of the change in solar irradiance is not due to solar activity, then probably the entire Sun is involved, either the convection zone where the Sun's energy is brought to the surface by turbulent gases, or perhaps even the nuclear furnace in the center of the Sun.

One way to help decide the importance of solar activity is to actually measure the "missing energy" from sunspots. Since sunspots can change quite a lot from day to day, observations are needed on at least a daily basis. Since it is a great deal of work to photometer all of the sunspots every day (there can be dozens at one time near solar maximum), some observatories take photographs of the whole disk of the Sun and then attempt to measure the area of each sunspot. Although easier and faster than photometry, this technique involves several major assumptions about the structure and similarities of sunspots. (Photographs are completely underexposed in the umbra of sunspots; all intensity information is lost.) It is usually assumed that all sunspots have the same average intensity. But what if some sunspots have larger umbrae than average, or cooler? Hoyt (1979) found evidence that the ratio of umbral to penumbral area in sunspots is a function of the phase of the eleven year solar cycle. If true, this variation would suggest that the solar convection zone changes its structure during the solar cycle. (The umbra is that central part of larger sunspots where the intensity is perhaps only 10 percent to 20 percent of the quiet Sun.) An important problem then is to measure the intensity of the umbra of a sunspot (in different color bands if possible). It must be kept in mind that light from the bright, surrounding photosphere will be scattered into the image of the sunspot. The elimination of this parasitic light is rather difficult but could be carried out with the help of another collaborator with a computer. The data needed to remove this parasitic light are the intensity of the aureole, that halo of light around the Sun. The intensity of the aureole will probably be in the range of 0.01 percent to 1 percent of the intensity of the center of the solar disk, depending on sky conditions, telescope optics, and wavelength of observation. Measurements of this aureole should be carried out before, during, and after sunspot measurements out to a distance of about one or two solar radii from the limb (15 to 30 arcminutes).

Lest I scare the reader away from this problem, keep in mind that the sunspot measurements are worthwhile even without correction for scattered light as long as the aureole measurements are made. The relative darkness of sunspots, uncorrected for scattered light, can still be used to intercompare sunspots to see if their umbral intensity is related to their area, age, or the phase of the 11-year solar cycle.

Observations of sunspots should be made with a narrow-band interference filter in the yellow or orange part of the spectrum. Wavelengths that are fairly clear of absorption lines, both in the photosphere as well as the sunspot should be chosen. At the San Fernando Observatory, measurements of sunspots are carried out with a Reticon diode array at a wavelength of 6264 Å. Other wavelengths may be more suitable for narrow-band filter photometry since at SFO, a spectrograph is used to isolate a 1.5 Å band. Our use of a diode array, a linear set of 512 discrete light sensors (as shown in Figure 5-5), permits us to obtain images of sunspots. This capability is useful for looking for brightness patterns around sunspots and minimizes telescope pointing problems. However, the collection of data must be carried out by computer, and the data stored on a high speed tape unit capable of recording 10^4 digital words per sec or more. Reconstructing these data into a picture for viewing requires a computer with a digital image display unit.

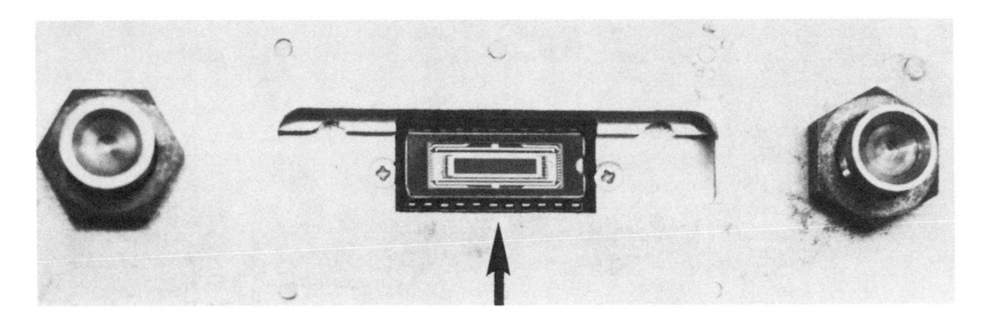

Figure 5-5. The Reticon 512 diode array mounted in the holder that replaces film in one channel of the vacuum spectroheliograph. This detector is used to photometer active regions, producing digital images with 512 x 512 pixels.

The serious amateur can obtain data on sunspots using a single aperture photometer. If the telescope has a large enough aperture, say 6 to 8 inches or larger, it should be possible to use a photometer aperture with a diameter of about 5 arcseconds and measure the umbral and penumbral intensity separately by letting the sunspot drift across the aperture. Letting the Sun continue to drift, one can obtain measurements of the scattered light aureole. For this work, a clock drive is not even required! If the seeing is poor or if the telescope has a small diameter, it may be better to use a large photometer aperture, say 40-60 arcseconds so as to include all of the sunspot. Again, the Sun can be allowed to drift across the aperture with the clock drive stopped.

III. EQUIPMENT

For solar photometry, one doesn't need a cumbersome, high gain, delicate photomultiplier tube (PMT). The best light sensor is a PIN photodiode. These devices are relatively inexpensive and rugged. A good source is United Detector Technology in Culver City, California, although other companies can also supply these devices. Most manufacturers also supply a model with an amplifier built into the device so that with a low power voltage supply, the detector combination will generate a signal voltage that is proportional to the light level. The advantages of these solid-state detectors is that they are quite linear in converting light level to an electrical signal and have a broad wavelength response, that of silicon, from about 4000 Å up to nearly 11,000 Å. A second amplifier will probably be required to boost the signal level up to a few volts. The data can then be recorded.

The best method of recording the data is to convert the signals to digital numbers and store these on magnetic tape or disk. Many microcomputers can be fitted with analog-to-digital converters (ADC) to accomplish the A-to-D conversion. Once converted, the data can be stored on whatever mass storage device happens to be hooked to the computer. One advantage of "going digital" is the ability to handle large amounts of data, yet retrieve them rapidly.

A chart recorder could be used to record the amplified signal but recording the scattered light aureole might require a logarithmic amplifier to have enough dynamic range to see the disk of the Sun, yet still record the aureole, whose intensity is lower than the disk by a factor of 10^3 to 10^4. Those with stellar photometers having gain controls in magnitudes may be able to use them as logarithmic amplifiers. A brightness change of 10^4 corresponds to 10 magnitudes. This range is approximately that needed to do photometry on stars as bright as Sirius and as faint as the fainter stars in the Bonner Durchmüsterung.

In order to reliably measure the light from a sunspot, there must be a way of viewing the Sun with the aperture superimposed. The simplest way might be to project an image of the Sun onto a screen having a hole cut out for the aperture. The light sensor could then be placed behind the hole; since silicon photodiodes are more uniform than PMT photocathodes, there is not as great a need for a Fabry lens as in the case of a conventional PMT photometer. A drawback to this projected image photometer is the additional scattering of light caused by the eyepiece itself. Nonetheless, this scheme is simple and should work, provided the eyepiece is of good quality and is clean. Probably a long focus eyepiece or an achromatic lens would suffice. A long focus lens would help to keep the eyepiece or projection lens from overheating. Schott Glass Works sells heat absorbing glass (as does Edmund Corporation), which could be placed before the projection lens. A somewhat better scheme optically is to let the direct solar image fall on a metal aperture. The light going through the aperture can then be filtered and reimaged onto the photodiode by the equivalent of a Fabry lens. This lens should be placed so as to form an image of the telescope objective on the photodiode to not lose any light and minimize the effects of flexure in the optical path.

An amateur astronomer can make a valuable contribution to this field because telescope time is so heavily committed at most observatories that photometry of sunspots is very seldom carried out.

Those possessing large telescopes, say 12 1/2-inch reflectors, adaptable to solar work, might consider the use of television in conjunction with video tape recorders to obtain photometric images of sunspots under good seeing conditions. There is at present some uncertainty concerning the structures that sometimes exist inside the umbra of a sunspot. Both dots and granules have been reported to exist in the umbrae of certain sunspots (Lawrence 1983). One requires the very largest telescope possible, 12-16 inches, of good optical figure. To control heating on secondary optics, one should consider a Gregorian configuration so most of the unused part of the image can be reflected away before striking the secondary mirror. The Solar Optical Telescope (SOT) is a 1.25 meter diameter Gregorian solar telescope to be built by NASA and flown on the Space Shuttle in the late 1980's or early 1990's. The San Fernando Observatory has a 24-inch vacuum reflector with a wide field of view, about 80 percent of the solar diameter, which requires a Cassegrainian optical system and results in considerable concentration of light on the Cassegrain secondary. Absorption of heat is minimized by the use of ultra-high reflective coatings. To prevent image distortion due to temperature changes, all mirrors are of "zero expansion" Cer-Vit. Other possible optical schemes can be devised to magnify the image and reduce the heating load on secondary optical components.

A well-made 12 1/2-inch reflector can approach a resolution of 1/2 arcsec. Some work on umbral dots, reported in the scientific literature, has been done with 8-inch telescopes (Lawrence, 1983). With a 12-inch telescope, even photography of umbral dots can be of interest (although a photograph has a fairly limited dynamic range.

An advantage of television is that it has a larger dynamic range than film (with ordinary developers). Because of continuous recording, moments of very good seeing can be singled out for later analysis. Eventually, a truly excellent TV recording could perhaps be digitized at a professional observatory or digital image lab for further analysis. As the price of A-to-D converters fall, the capability of digitizing select video frames will come within the grasp of the private citizen. Solid-state TV cameras should be more rugged and will have better dimensional stability than vidicons. Solid-state cameras will probably have a broader spectral response as well.

IV. SUNSPOT PHOTOMETRY

With sunspot photometry, at whatever spatial resolution is attainable, sunspot "drift curves" are of value to show the average intensity in a sunspot relative to the photosphere. A question to be answered is, how does the intensity and size of a sunspot, especially the umbra, change with the age of the sunspot?

A useful adjunct to photometric drift scans is a good photograph of the spot. Since sunspots do have internal structure such as umbral dots, light bridges, etc., it would be helpful to know what structures may have been crossed by the photometer aperture.

A simple flip mirror could allow the image to be switched from the photometer to a mounted 35 mm camera. (Be sure that the light flux is reduced to a safe level so that the camera shutter is not damaged. A dichroic mirror to throw away most of the energy before the camera is helpful.)

V. DRIFT SCAN PHOTOMETRY

A question of major importance to solar astronomy as well as climatologists is whether the output of energy from the quiet Sun (that is that part of the Sun uninvolved with magnetic fields) changes with time. In other words, if solar activity, sunspots, and faculae were not present, would the Sun be variable in its output? If it is variable, what is the time scale of the variation--11 years, 22 years, or longer?

One simple method to determine if the quiet Sun is or is not variable is to make drift scans. Simply place the aperture of a photometer in the sky several arcminutes west of the west limb of the Sun and let the Sun drift through the aperture. This technique is especially useful if one has no high speed electronic recorders and if the sky does not change its transparency or scattering rapidly. Drift scans obtained at the San Fernando Observatory with the 28 cm vacuum telescope are shown in Figure 5-6.

These scans were obtained with a PIN diode as part of a UDT mod. 161 photometer with a green filter, Schott VG-14. The scans were recorded using a Clevite/Brush recorder. The effective wavelength of the combination is about $\lambda \approx 5300$ Å. The aperture was approximately 2mm which corresponds to 37 arcsec on the sky (see Figure 5-7). The optical set-up is shown in Figures 5-8 and 5-9. The drive on the 61/28 cm vacuum telescope was turned off to permit the drift scan.

If this technique of measuring the solar limb darkening is so simple, why haven't more observations of this nature been obtained? Partly, the answer is that a great deal of the professional solar astronomers' attention has been directed at space observations of solar flares. As a consequence, many ground-based observatories became involved in observations of solar activity in support of the space observations. Space-based solar astronomy had two peaks, the first being the Skylab era, from about 1973 until about 1974 or 1975. Actual manned operations began in June 1973 and ended in January 1974. The second peak was the Solar Maximum Mission, a spacecraft and research program that was devoted to studying the Sun at sunspot maximum, when flares are most frequent. Again, considerable effort was put into ground-based observations of solar activity to provide data for the SMM that could not be obtained from the spacecraft, such as high resolution H_α filtergrams, magnetograms that map the strong solar magnetic field, and other useful data that may help in determining how the energy of the flare builds up and is stored, and what triggers its sudden energy release.

Figure 5-7. The silicon detector head of the United Detector Technology photometer used to obtain the drift scan of Figure 5-6 showing the 2mm "pinhole" in an aluminum baffle.

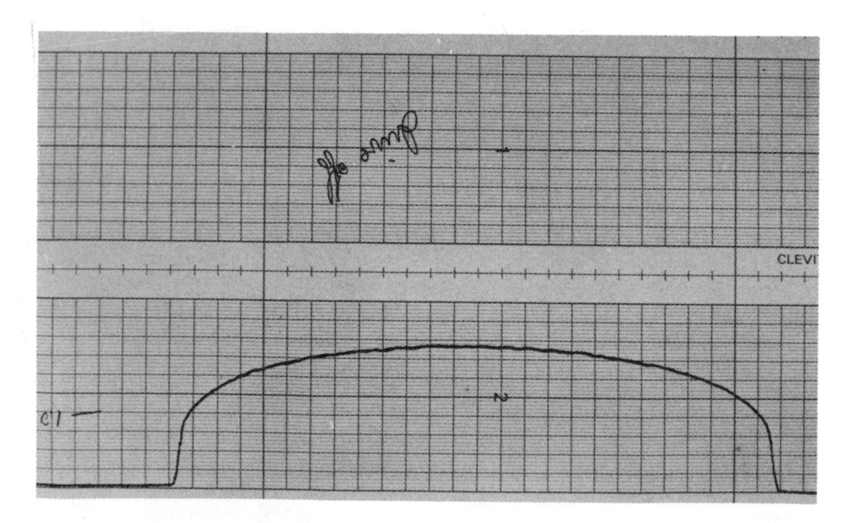

Figure 5-6. A drift scan of the quiet-sun limb darkening through a Schott VG14 green filter with a passband of about 800 Å centered near 5300 Å. Most of the small wiggles in the scan are real, caused by short-lived convection cells. Many such scans should be averaged together to get an accurate global limb darkening.

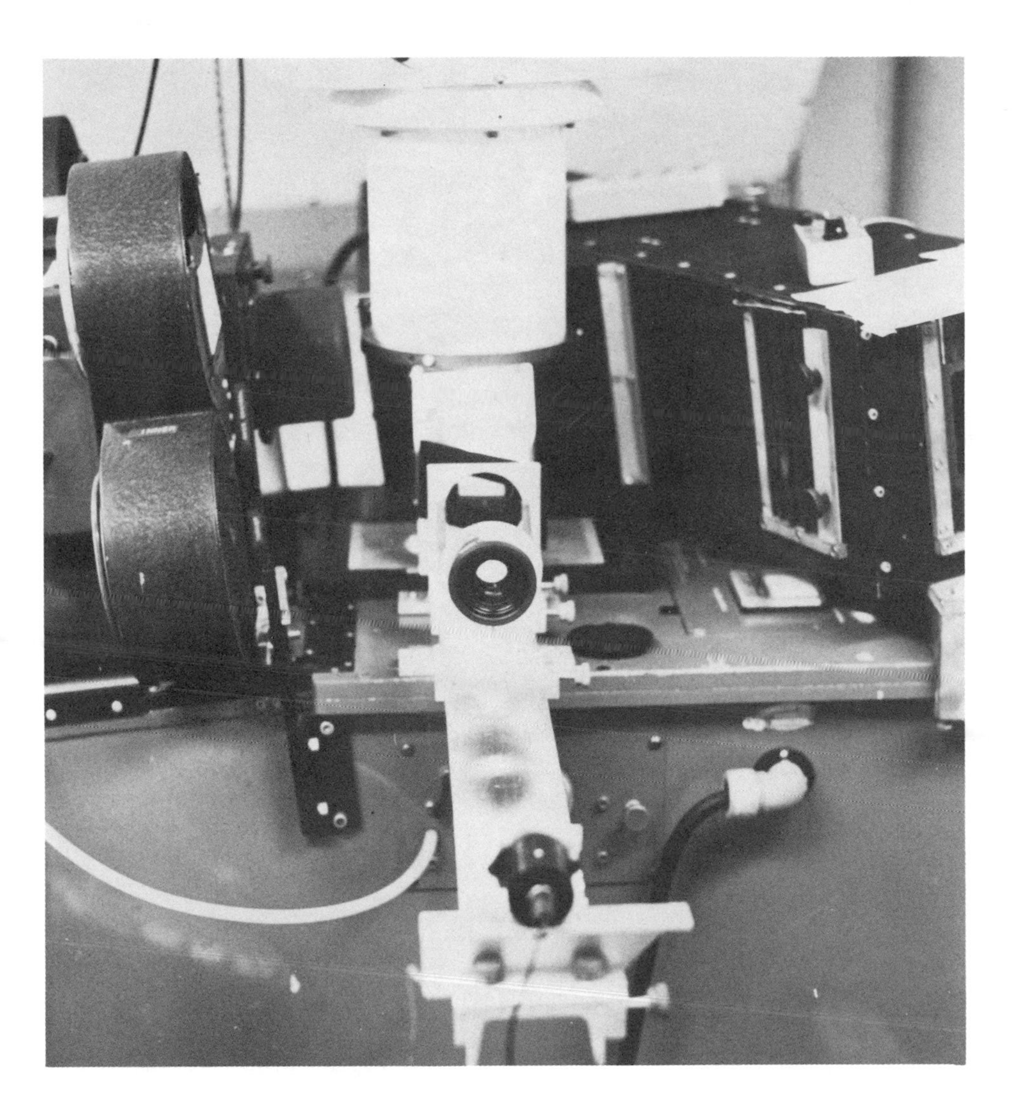

Figure 5-8. The optical bench, fed by the 28 cm vacuum telescope for making drift scans of limb darkening. The optical bench is bolted to the vacuum spectroheliograph.

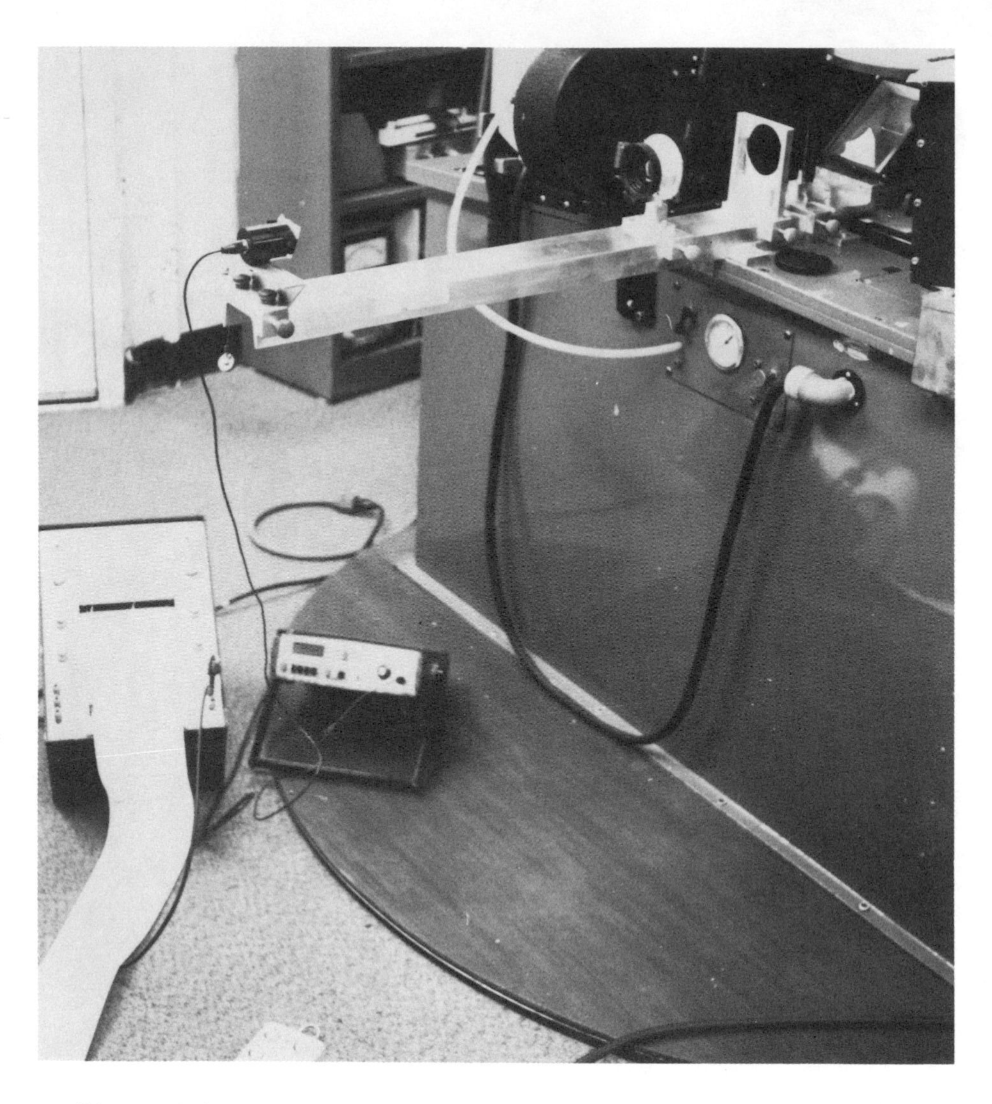

Figure 5-9. The set-up for the drift-scan of Figure 5-6 showing the UDT photometer connected to a Gould/Brush recorder.

Those solar astronomers who have not been involved in the space program, either directly or indirectly, have been scattered among a variety of problems, chief among them being the subject of solar seismology, i.e., the study of the solar interior from oscillations observed in the solar atmosphere (Eddy 1978). In brief, not very many solar astronomers have been studying solar variability, at least not in terms of quiet Sun limb darkening variability.

Why would we expect that the limb darkening might change? There are at least two mechanisms that could cause a change in the limb darkening. One is a change in convective efficiency. The outer part of the Sun is so opaque that radiation can no longer carry the flow of energy. As a result, the solar gases themselves must move upward, carrying the heat as an increased gas temperature. The gases reach the photosphere where they can "feel" the presence of cold outer space, whereupon the gas cools by radiation and flows back into the Sun to repeat the cycle. This region is called the convection zone and is thought to extend down into the Sun about 160,000 km below the surface or about 23 percent of the solar radius (Gilliland 1982).

A change in the efficiency of convection, due to the presence of subsurface magnetic fields, for example, can lead to a change in the solar luminosity and an associated change in the limb darkening of the quiet Sun. Also, the radius of the Sun should change, although current attempts to calculate the change in solar radius for a given change in luminosity, are not in agreement and in some cases even the sign of the change is not agreed to by various authors (Gilliland 1982).

Among observations which strongly suggest that something is changing with the solar cycle are those of Livingston and Holweger (1982). They report finding an increase in the efficiency of convection in the outer 100 km of the solar atmosphere based on measurements of the strength of several spectrum lines between the years 1976 to 1980. This period corresponds to the time of solar minimum to solar maximum. If their observational result is indeed caused by a change in convection in the outer layers and not by unresolved faculae, then the quiet Sun limb darkening should show a noticeable change. Clearly, reliable observations of limb darkening will help to check the findings of Livingston and Holweger.

VI. PRESENT AND FUTURE OBSERVATIONS

Photometry of the Sun at the San Fernando Observatory is presently being carried out with the Reticon linear diode array, used on the vacuum spectroheliograph, and with the Extreme Limb Photometer (ELP). The diode array produces digital images of active regions, sunspots, and faculae, that have 512 rows and 512 columns of pixels as shown in Figure 5-10. Thus, a diode array picture will have approximately a quarter of a million numbers, each one in the range of 0 to 4095 decimal. The larger the number, the brighter the Sun was at that point in the picture. To obtain the data for such a picture requires a computer with an analog-to-digital converter (ADC) and a digital tape recorder, as mentioned earlier. To display such a picture requires a digital image processor. The SFO has a Quantex digital image memory/processor controlled by a Varian/Sperry Univac 620i minicomputer as shown in Figure 5-11. This unit also has front panel controls for manually controlling the picture display. It can digitize a live or recorded video image and store that data in memory. Only one picture at a time can be held in memory. To store processed data for later use, the contents of the Quantex memory can be extracted by the main computer, though the "IEEE bus," and written onto nine or seven track magnetic tape. This digital tape then contains one or more digital pictures, as tens of thousands of binary numbers, for later processing by a larger computer.

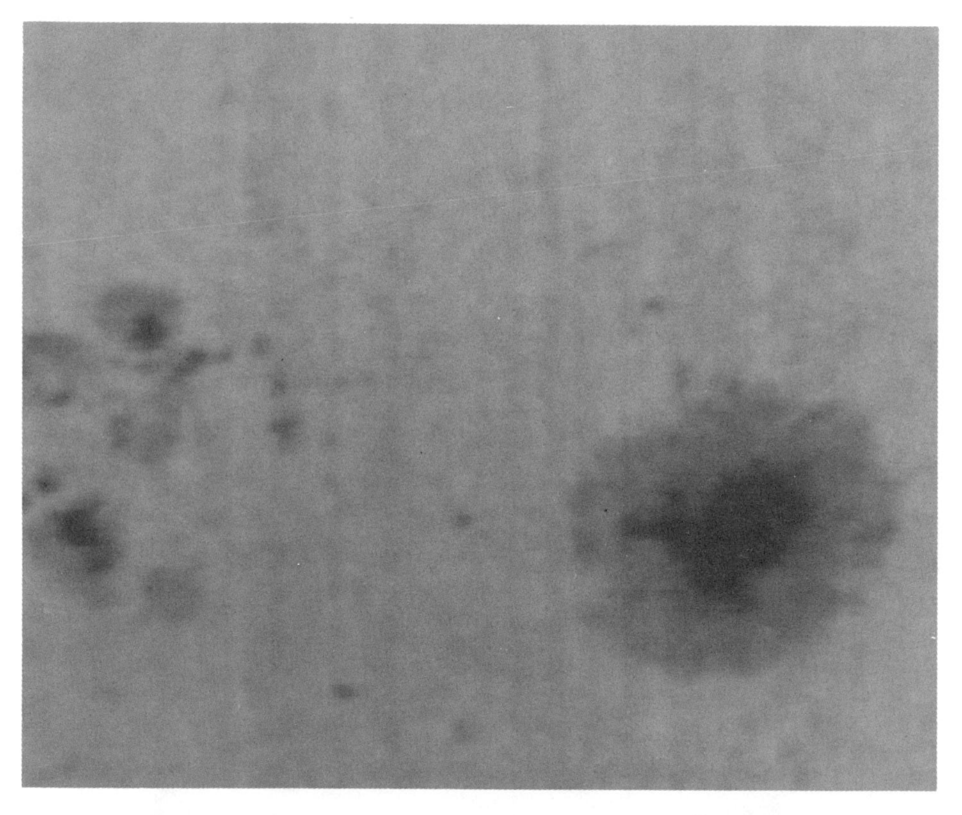

Figure 5-10. Digital image created from 512 x 512 diode array. Shown is an active sunspot region.

Figure 5-11. The SFO digital image system for the processing and display of digital images from magnetic tape or analog video signals, live or recorded. The Quantex is the black box in the rightmost equipment rack.

Obviously, this type of data offers great versatility in analysis but requires a great deal of equipment in order to handle it. A reduction in the volume of data required for scanning, photometrically, an active region is achieved with the Extreme Limb Photometer (ELP). This device, with its 39 arcsecond long slit is able to photometer a solar active region in one or two swaths, each swath requiring about 10 seconds of scanning time and each corresponding to a slightly different position of the telescope. If the active region is large, several swaths may be required in order to cover the entire region.

The ELP was designed to scan active regions near the solar limb, thus its circular scanning design. The ELP rotates its 3 arcsecond wide by 39 arcsecond long slit around a circular path once a second of time as shown in Figure 5-12. When studying solar activity near the limb of the Sun, the photometer and telescope are pointed at the center of the solar disk. With this pointing, the slit travels on or near the solar limb. The quiet sun gives a fairly steady signal with faculae and sunspots producing higher or lower brightness, respectively, from the quiet Sun. The scanning pattern (circular) slightly complicates the data analysis for regions not on the limb, but the mechanical simplicity more than compensates for the circular scan because the scanning mechanism is just a stepper motor and two precision spur gears.

Figure 5-12. A photograph of the Extreme Limb Photometer (ELP). This photometer drives a 3 by 39 arcsecond slit in a circular path across solar active regions.

The diode array has a spatial resolution determined by the one arcsecond square size of the individual diodes and the width of the entrance slit of the spectroheliograph. If we wish to photometer a 40 arcsecond square region on the Sun, we must digitize and store 1600 pixels. With the ELP, however, this same area can be covered by $40/3 \approx 13$ data points, each one corresponding to a particular position of the ELP slit. Although we no longer can obtain a "picture" of the active region, we save a considerable amount of data processing.

Since the circular scan of the ELP has to be placed over the active region to be studied, the intensity of the quiet Sun changes from place-to-place in the scan. This is because the ELP had to be pointed away from disk center if anything other than the limb is to be scanned photoelectrically. It would be much better if the scanner did not have to be shifted away from disk center in order to scan an arbitrary active region. With this difficulty in mind, we have designed a Full Disk Photometer (FDP) and we are seeking funds to construct and operate the FDP.

The FDP will be like the Extreme Limb Photometer in that it will be a rotary scanning photometer. Instead of having a single detector, however, and requiring the telescope to be off-pointed from disk center to scan an active region, the FDP will remain Sun-centered at all times (except for calibration scans). The detector will be a linear diode array having 512 detectors aligned along the solar radius. The diode array will be rotated by a stepper motor so that the diode array will sweep over the whole solar disk in about 1-1/4 minutes. At the end of a complete rotation, the stepper motor will return the entire assembly back to its starting position to repeat the scan, possibly at another wavelength. One complete rotation will consist of 3220 steps, each step corresponding to a tangential motion of two arcseconds at the limb. Data will be averaged after being digitized (at about 21,000 pixels per second) and the averaged picture written to magnetic tape or magnetic disk storage.

The image of the Sun, if unprocessed, would contain 512 x 3220 or 1.6 million pixels, each pixel represented by a 12 to 16 bit binary number. In operation, the data will be averaged to reduce the storage requirements. We estimate that the device should be able to detect a variation in solar irradiance in the midvisible of about 100 parts per million relative to the unperturbed, quiet Sun. The FDP will also be used to detect changes in the limb darkening of the quiet Sun.

Since high spatial resolution is not important, we anticipate using a 15-20cm (6-8 inch) diameter lens as the objective for this experiment, thus permitting the use of the 61 cm vacuum telescope for high resolution studies of the magnetic fields of sunspots and faculae (see Figure 5-13).

As the solar activity minimum of 1986-1987 approaches, we expect to be able to determine the quiet Sun limb darkening with greater ease than would be the case with many sunspots around. As the solar minimum approaches, sunspots will more often occur singly, thus permitting a clear identification of any irradiance fluctuation, seen from space or the ground, with that isolated sunspot.

A proposal, with Dr. Adrian Herzog, to build and operate this full disk photometer has been submitted to the federal government and will be reviewed. We are hopeful that this next logical step in solar photometry can begin before long at the San Fernando Observatory.

Figure 5-13. The 9-inch folded refractor on its guided spar. This telescope will feed the Full Disk Photometer when completed.

ACKNOWLEDGMENTS

I wish to acknowledge the contribution of several individuals to some of the results shown here. They are Drs. John Lawrence and Chris Shelton and Michelle Hinnrichs for work leading to the diode array images, Bill Mott for mechanical construction, and Scott Templer for help with the figures of Chapter 5. Figure 5-1 was from a photograph obtained by Steve Thorman.

REFERENCES

Allen, C. W., 1973. Astrophysical Quantities, Athlone Press, London.

Chapman, G. A., and Klabunde, D. P., 1982. Astrophysical Journal, 261, 387.

Eddy, J. A., ed., 1978. The New Solar Physics, Westview Press, Boulder.

Frazier, Kendrick, 1982. Our Turbulent Sun, Prentice-Hall, Englewood Cliffs.

Gibson, E. G., 1973. The Quiet Sun, NASA SP-303, U. S. Government Printing Office, Washington D.C.

Gilliland, R. L., 1982. Astrophysical Journal, 253, 399.

Hoyt, D. V., 1979. Climate Change 2, 79.

Lawrence, J. K., 1983. "The Spatial Distribution of Umbral Dots and Granules," Solar Physics, in press.

Livingston, W., and Holweger, H., 1982. Astrophysical Journal 252, 375.

Mayfield, E. B., Vrabec, R., Rogers, E. H., Janssens, T. J., and Becker, R. A., 1969. Sky and Telescope, 27, 208.

Mitton, Simon, 1981. Daytime Star: The Story of our Sun, Charles Scribner's Sons, Boston.

Noyes, R. W., 1982. The Sun, our Star, Harvard University Press, Cambridge.

Rosen, W. A., Foukal, P. V., Kurucz, R. L., and Pierce, A. K., 1982. Astrophysical Journal (Letters), 253, L89.

Sofia, S., ed., 1981. Variations of the Solar Constant, NASA Conference Publication No. 2191, Goddard Space Flight Center, Greenbelt.

White, O. R., ed., 1977. The Solar Output and its Variation, University of Colorado Press, Boulder.

Willson, R. C., Gulkis, S., Janssen, M., Hudson, H. S., and Chapman, G. A., 1981. Science, 211, 700.

6. LOW SPEED EQUIPMENT

Jeffrey L. Hopkins

I. INTRODUCTION

The basic photoelectric photometry system consists of the telescope, photometer head, and supporting electronics (see Figure 6-1). The telescope can be either a refractor or reflector. The refractor will have a disadvantage if filter photometry is to be done, particularly in the ultraviolet region. Considering cost and ease of use the Cassegrain is an excellent choice. A unique characteristic of photoelectric photometry is that the quality of the optics is not a critical factor.

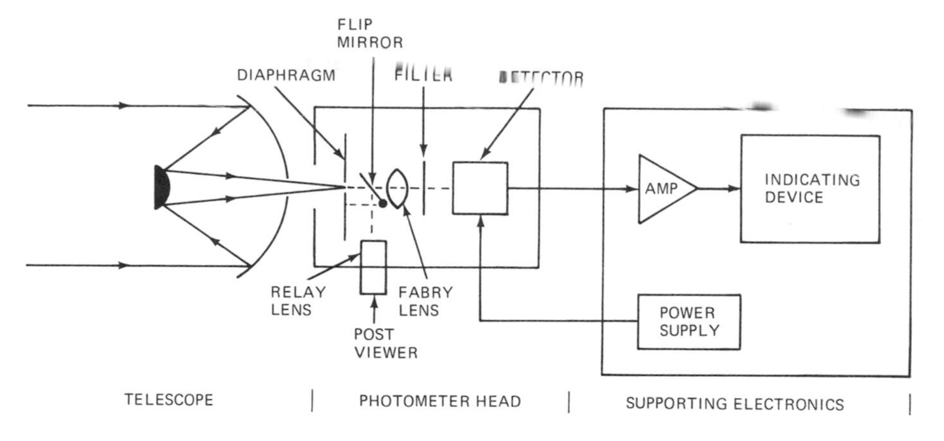

Figure 6-1. Basic photoelectric photometer system.

The photometer head is mounted on the telescope such that the focal plane falls on the diaphragm opening of the photometer head. Several selectable diaphragm sizes allow masking of the sky around the object to be measured. A flip mirror is used to divert the light to a post-viewer eyepiece where the image can be centered in the diaphragm opening. A Fabry lens is used to minimize the effects of movement of the image on the detector. Several selectable filters are placed before the detector to allow filter photometry. The detector may be either solid state or a photomultiplier tube (PMT). The solid-state detector has advantages of being virtually indestructible from excessive light, requiring only a low voltage power supply (batteries can be used), being very rugged mechanically, and being sensitive in the infrared region. Its disadvantages are that it is not as sensitive as the PMT in the visual through ultraviolet region, can not (as of this writing) be used as a photon counter and generally has a poorer signal-to-noise (S/N) ratio than the PMT. The PMT has been around for many years and has proved very useful and reliable. Its disadvantages are that it requires a high voltage power supply (-1000VDC @1mA), it is not very sensitive in the infrared region, it is fragile (glass tube) and can be damaged by excessive light. Its advantages are that it is extremely sensitive, it can be used with either analog or photon counting systems (and is therefore capable of very good S/N and dynamic range), and it has been a standard for many observatories doing photometry.

The supporting electronics consists of an amplifier and some means of indicating the resulting signal. With analog systems the amplifier is used to amplify an average DC current (typically on the order of nano-amperes, 10^{-9}A). The resulting signal is then displayed on a meter (analog or digital) and/or chart recorder. It is also possible to convert the amplified signal to a frequency with a voltage-to-frequency converter. The resulting frequency can then be displayed on a frequency counter and/or read by a computer. It must be remembered that this is still an analog and not a photon counting system. It still lacks the dynamic range (the amplifier gain must still be switched) and has a poorer S/N.

With photon or pulse counting systems the actual pulse that results from a photon striking the detector is counted. The pulse is typically of millivolt level and sub-nano second pulse width (when terminated in 50 ohms). A high-speed amplifier is used to amplify the pulse level. A comparator (discriminator) then is used to compare the amplified pulse level with a preset threshold level. Any pulses exceeding the threshold level cause the comparator to produce a trigger pulse. The trigger pulse causes a mono-stable (one shot) to produce a pulse of predetermined pulse width and amplitude for each triggering pulse. One problem is that if the pulses are coming too fast, the one shot can not start another pulse until the first pulse is complete plus a small amount of reset time. The time for which no more pulses will be counted is called dead time. For faint objects this is not much of a problem, however the brighter the object, the more pulses per second are produced by the PMT. A calibration can be performed and the dead time for a system can be calculated and allowed for. The one shot output can then be sent to a frequency counter and the pulses per second displayed. For faint objects the gate time can be increased for a more accurate count. The pulses can also be counted with a digital counter and the results fed directly into a computer. The price of a complete photon counting system can be less than the price of a chart recorder so the photon counting system is relatively economical.

II. COMMERCIAL EQUIPMENT

As of this writing, Optec, Inc., 199 Smith Street, Lowell, MI 49331, is the only producer of solid-state photoelectric photometry equipment. They have several models to choose from including a completely self-contained battery operated unit. The units are well-designed and well-built. If portability, price, or near IR operation are considerations, then the Optec solid-state photometer is worth looking into.

EMI Gencom Inc., 80 Express Street, Plainview, NY 11803, has been producing photometry equipment for high energy physics for many years. Recently, they have introduced the STARLIGHT-1 Photon Counting Stellar Photometer. The unit consists of a large photometer head with an end-on PMT, UBV filters, 6 diaphragm sizes, and high voltage power supply. The supporting electronics is a 10 MHz universal counter driven by a LeCroy MVL100 (which is located in the head). Besides having an optional remote control for the counter, the unit has a capability to be interfaced with a computer. An 8-digit LED readout is provided on the counter. The unit is an excellent choice for a large telescope where size and weight of the head would not cause problems with balance and clearance.

Hopkins Phoenix Observatory, 7812 West Clayton Drive, Phoenix, AZ 85033, offers a compact lightweight photometer head for under $500. It is designed with the small telescope in mind such as the Celestron C-8. It is under two pounds, easily clears the C-8 fork mount, contains UBV filters plus a dark slide, and four selectable diaphragm sizes. Supporting electronics consists of two units. An adjustable high voltage power supply with a 3 1/2-digit LED digital panel meter, and a 10 MHz universal counter with 8-digit LED readout and remote control unit for the counter. In addition, the counter unit has a TTL output for interfacing with a computer and a X10 output for analog measurement (such as monitoring the signal with a chart recorder. The complete system provides the small observatory with photon counting equipment for under $1000.

It is suggested that any interest in the above equipment be pursued by writing directly to the company. An alternative to purchasing a commercial unit is to build your own. Besides saving money, this approach allows the builder to explore some of the finer areas of photoelectric photometry.

III. BUILDING THE PHOTOMETER HEAD

Perhaps the most difficult part of a PEP system is the design and construction of the PEP head. It must be light tight, operate easily and smoothly (so as not to disturb the telescope), be properly aligned, and with small telescopes be small and light enough so as not to seriously unbalance the telescope or restrict the telescope's operation. A basic head is shown in Figure 6-2. To save space, configurations can be used with fixed mirrors to allow the PMT to be mounted in positions other than directly in line with the optical axis. Diaphragms and filters can be mounted as shown in Figure 6-3, or on the more conventional slides.

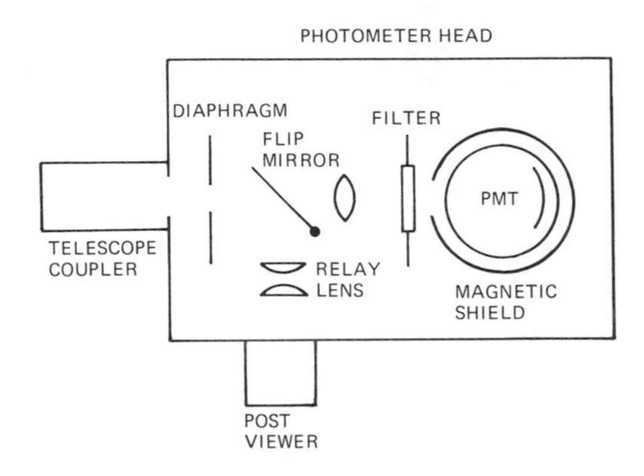

Figure 6-2. Basic photometer head.

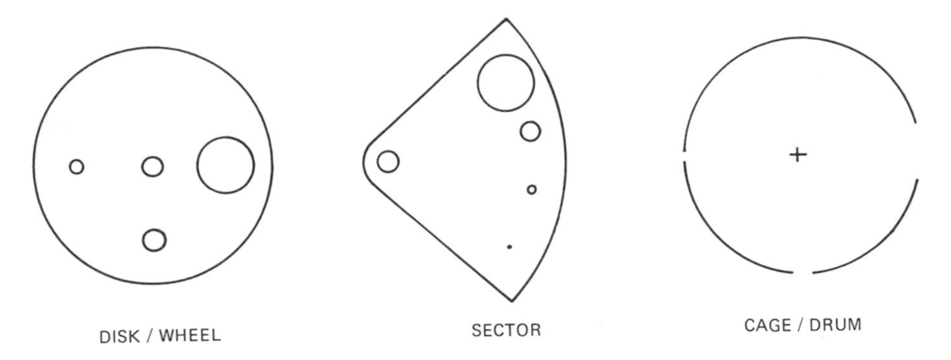

Figure 6-3. Diaphragm/filter mountings.

The choice of the PMT is important. Depending upon the application and cost, an end-on or side window type PMT might be selected. The end-on type have advantages but are more expensive and difficult to design the PEP head around. To avoid problems with magnetic fields (even the earth's), a magnetic shield should enclose the PMT. Usually the same manufacturers of the PMT's can supply a magnetic shield for the different types of tubes.

IV. BUILDING SUPPORT ELECTRONICS

The supporting electronics consists of power supplies (low voltage and high voltage for PMT's), an amplifier and some means of indicating or recording the resulting signal. With analog systems, the amplifier is used to amplify an average DC current (typically in nanoamperes, 10^{-9}A). The amplified current (or voltage across a resistance) is then displayed on a meter (analog or digital) and/or chart recorder. By using a voltage-to-frequency (V/F) converter the amplified voltage can be converted to a frequency which is proportional to the voltage. This frequency can then be displayed on a frequency counter and/or interfaced to a computer for direct entry into the computer.

With photon or pulse counting systems the actual pulse that results from a single photon striking the photocathode is counted. A fast (high frequency) amplifier is needed to amplify the resulting pulse, typically 100 MHz or more. High frequency techniques must be used when working with photon counting, such as shielding, ground planes, and proper terminations. There are no problems with leakage currents as with analog systems and the dynamic range is limited only by the counter (no gain switching required). The S/N is better than with the analog systems and the price of a photon counting system can be less than a typical chart recorder. A typical photon counting system (see Figure 6-4) consists of an amplifier, comparator, mono-stable (one shot), and counter. The small pulse from the PMT is amplified and sent to a comparator. If the pulse amplitude exceeds a predetermined threshold level, the comparator produces an output which triggers the mono-stable. The mono-stable has an external timing circuit which determines the pulse width of the output pulse. Thus, for every (except for dead time and pulses below the threshold) input pulse from the PMT the mono-stable produces a pulse of predetermined amplitude and duration. The LeCroy MVL100 combines all these functions into one 16-pin DIP device.

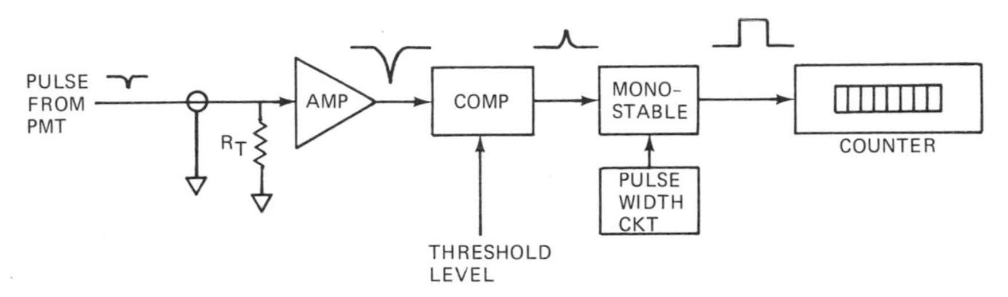

Figure 6-4. Photon/pulse counting system.

Low Voltage Power Supply

Since the introduction of the three terminal voltage regulator, the design of low voltage power supplies has become relatively easy. Two very important considerations that are sometimes overlooked are the power dissipation/heat sink requirements and adequate input voltage to the voltage regulator. To provide proper load and line regulation, the voltage regulator should have at least 3-volt differential between the input to the regulator and the output. This means for a 5-volt regulator the input must be at least 8-volts. Many people try to use 6.3 VAC transformer to build 5-volt power supplies. While this will work, it will not provide adequate regulation. A typical low voltage power supply is shown in Figure 6-5.

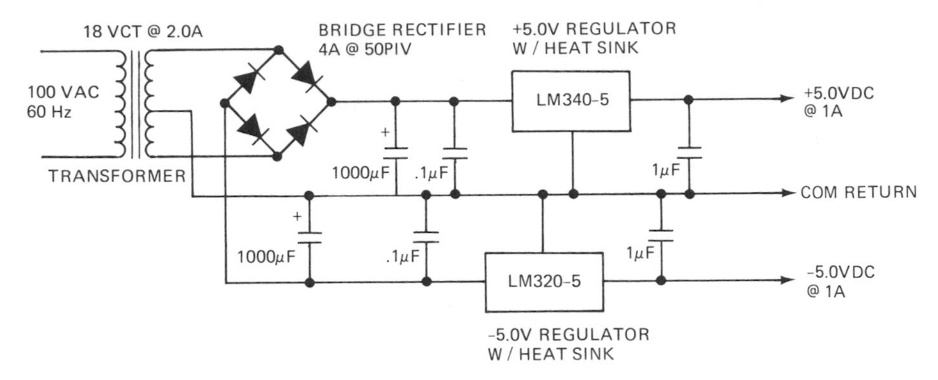

Figure 6-5. +/- 5 VDC Complementary power supply.

The second important consideration is the power dissipation of the voltage regulator. The voltage drop across the voltage regulator times the maximum current through the regulator gives the power dissipation. For a LM340-5 in a TO-220 package, the power dissipation is rated at 15 watts maximum. It should be noted that 15 watts is with an infinite heat sink or with a high rate of air flow over the heat sink. For most applications with a moderate heat sink and air flow the power dissipation should be figured at 7.5 watts maximum. For a 3-volt drop that means a maximum 2.5 amperes. However, the LM340-5 is rated at 1.5 amperes maximum so that will be the limiting current.

High Voltage Power Supplies

The use of a PMT as a detector requires a high voltage power supply. A 1P21 PMT requires from -500 VDC to -1000 VDC at up to 1 mA of current (the value of current depends mainly on the voltage divider design). The power supply should be regulated to ± 0.1% or better. Since the load-to-no-load current change with a PMT is typically in the 10^{-9}A range and the voltage divider (which is a constant load) is a factor of 1000 times greater, the load regulation becomes relatively unimportant. Thus, by providing good line regulation the power supply should be sufficiently regulated. Two different high voltage power supply designs are discussed. Each is easily built and provides reliable operation.

By using a three terminal adjustable voltage regulator, the HPO design shown in Figure 6-6 allows line regulated high voltage of -300 VDC to -1000 VDC at 1 mA. The CA555 is a timer IC used as an oscillator. The timing circuit allows the frequency of the CA555 to be varied from 10 kHz to 70 kHz. This allows a fine tuning for optimum operation. The square wave output of the CA555 turns the drive transistor (2N2219) on and off. This pulsing of the primary of the pot core transformer produces a stepped up voltage on the secondary of the transformer. There is approximately a 100:1 step-up so 10 volts on the primary will produce 1000 volts on the secondary. In actual practice a voltage of 7 to 8 volts on the primary will produce 1000 volts on the secondary. A calibration pot (100k ohms and 10 turns) is used to allow the high voltage to be measured with a low voltage meter. A high impedance (10 megohms or greater) meter should be used otherwise the meter will load the circuit. An alternate approach is to use a 0 to 1 mA meter in series with a 1 meg ohm resistor.

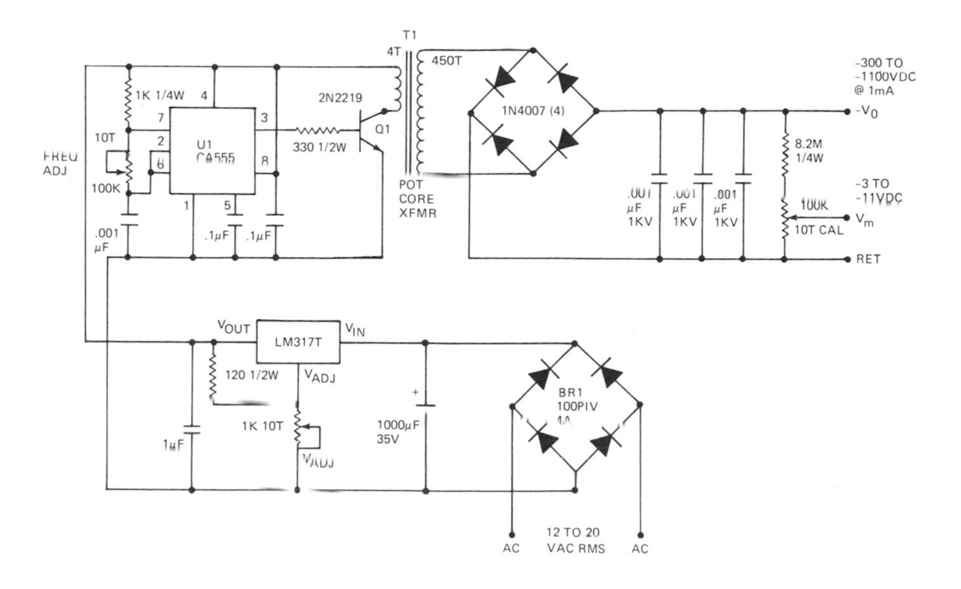

Figure 6-6. HPO high voltage power supply.

Louis Boyd of Phoenix, Arizona has designed the very stable high voltage power supply shown in Figure 6-7. A voltage feedback is compared with a reference voltage generated by the AD581 band-gap reference device. Variations in the output voltage are then corrected by the error amplifier. The power supply provides excellent regulation.

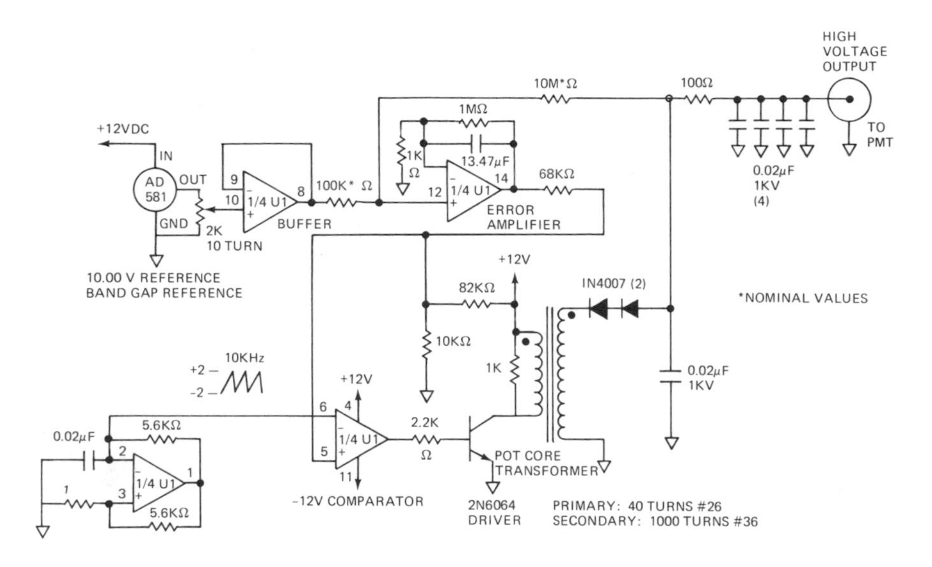

Figure 6-7. Louis Boyd's regulated PMT high voltage power supply.

The LeCroy MVL100 Low-Level Amplifier Comparator

The MVL100 (shown in Figure 6-8) is a high speed monolithic device intended for low-level pulse counting. It offers exceptionally low threshold, typically ± 200 μV and time slewing of 1.5 nS from 2X threshold to 20X threshold. Both inputs and outputs are fully differential and compatible with twisted pair signal transmission. The threshold is adjustable from less than ± 250 μV to greater than ± 3.2 mV. Output duration is variable (10 nS to 10000 nS) via external trim. The input circuit provides an impedance of greater than 5000 ohms. Input protection to ± 2 kV is provided to prevent damage from high voltage transients. The outputs of the MVL100 are standard emitter coupled logic (ECL) levels (-0.8V to -1.6V). The X10 output is an analog output. It can drive 50 Ohms via a coupling capacitor. A 390 ohm resistor pullup to + 5VDC improves the linearity. If not used, this output can be left open. The threshold voltage can be adjusted from -1 VDC to -16 VDC. The effective input threshold is a linear function of the threshold voltage with a slope of 200 μV/V. For further information about the MVL100 write to: LeCroy Research Systems Corp., 700 South Main Street, Springvalley, NY 10977.

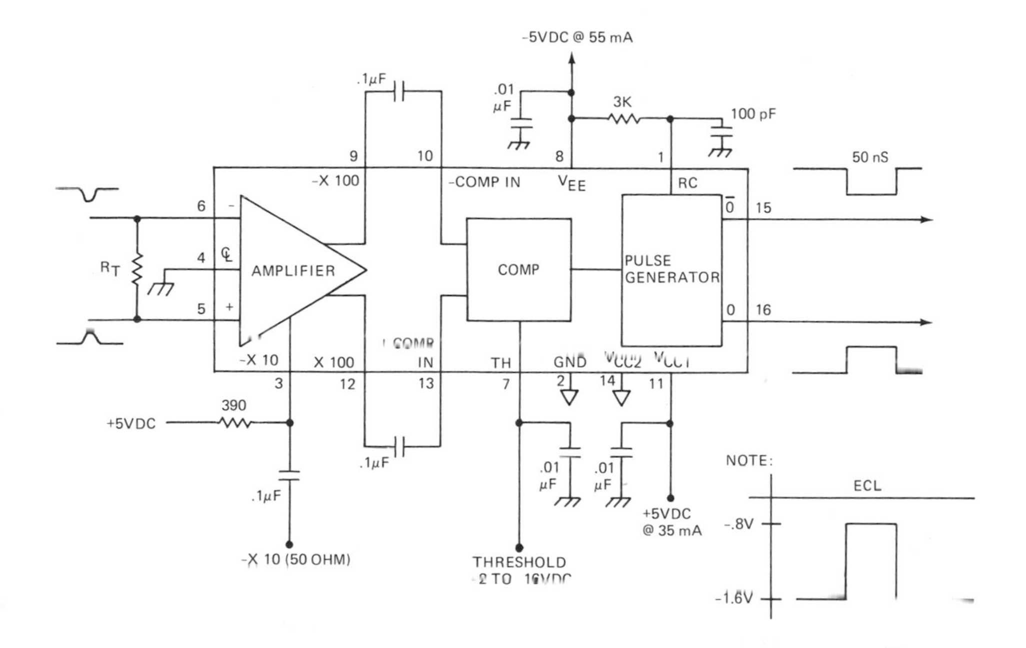

Figure 6-8. MVL100 low-level amplifier/comparator.

The MC10125 Quad MECL-to-TTL Translator

The MC10125 shown in Figure 6-9 is packaged in a 16-pin DIP. It is a quad translator designed to interface data and control signals between MECL (ECL) and saturated logic sections (TTL) of digital systems. The MC10125 incorporates differential inputs and Schottky TTL "totem pole" outputs. Differential inputs allow use as an inverting/non-inverting translator or as a differential line receiver. The V_{BB} reference voltage is available on pin 1 for use in single ended input biasing. The outputs go to a low logic level whenever the inputs are left floating. For additional information about the MC10125 write to: Motorola Semiconductor Products, Inc., Box 20912, Phoenix, AZ 85036.

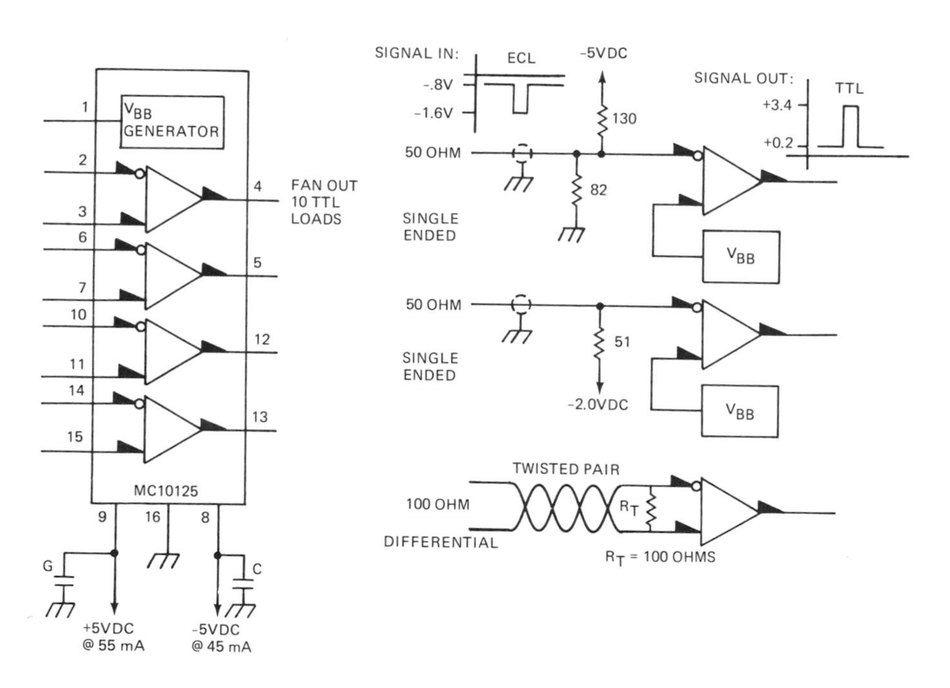

Figure 6-9. MC10125 MECL to TTL translator.

Using the MVL100 and MC10125 for Photon Counting

Figure 6-10 shows a typical circuit for using the MVL100 and MC10125 for translating the PMT output pulses into TTL pulses for use with a TTL input counter. Two outputs are used on the MC10125. One is intended to go to a counter and display while the other one can go to a TTL counter input of a computer interface.

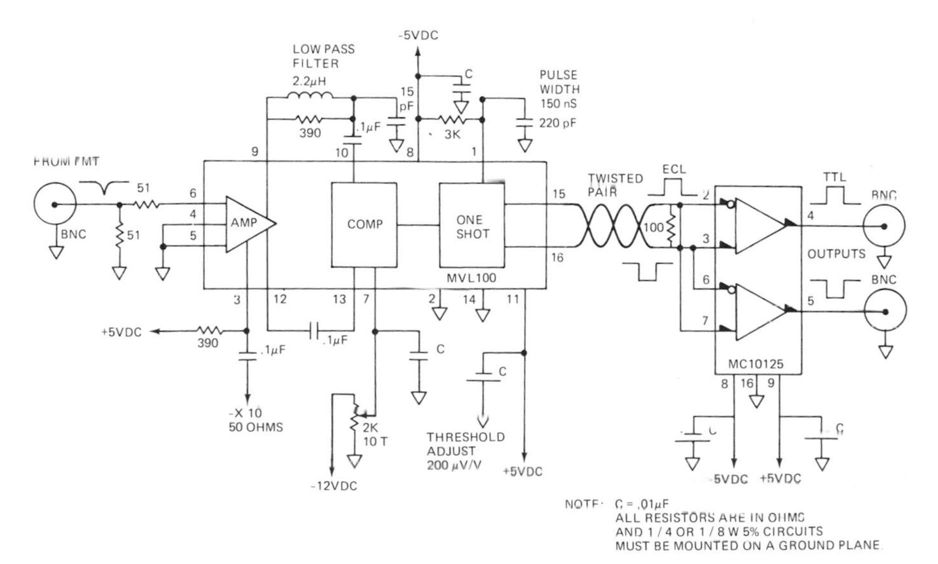

Figure 6-10. MVL100/MC10125

The Intersil ICM7226A 10 MHZ Universal Counter

The ICM7226A series of counters were the first IC's to contain all the active circuitry needed to implement a universal counter on a single chip. The ICM7226A, shown in Figure 6-11, is housed in a 40-pin DIP package. The device is available for $32 and an evaluation kit is available for $75 complete except for a +5VDC power supply. The ICM7226A is a fully integrated Universal Counter and LED display driver. It combines a high frequency oscillator, a decade timebase counter, an 8 decade data counter and latches, a 7 segment decoder, digit multiplexer and 8 segment and 8 digit drivers which can directly drive large LED displays. The counter inputs accept a maximum frequency of 10 MHz in frequency and unit count modes and 2 MHz in other modes. Both inputs are digital inputs. In many applications amplification and level shifting will be required to obtain proper digital signals for these inputs. The selectable features are given in Table 6-1.

TABLE 6-1

Functions Selected by each Digit for these Inputs

	Function	Digit
Function Input Pin 4	Frequency	D_0
	Period	D_7
	Frequency Ratio	D_1
	Time Interval	D_4
	Unit Count	D_3
	Oscillator Frequency	D_2
Range Input Pin 21	.01 Sec/1 Cycle	D_0
	.1 Sec/10 Cycles	D_1
	1 Sec/100 Cycles	D_2
	10 Sec/1K Cycles	D_3

Note that the ICM7226A can be triggered into destructive latchup if either of the inputs are applied before the +5 VDC is applied to the ICM7226A or if the inputs or outputs are forced to voltages exceeding +5 VDC and ground by more than 0.3 volts. For more information about the ICM7226A write to: Intersil, Inc., 10710 N. Tantau Avenue, Cupertino, CA 95014.

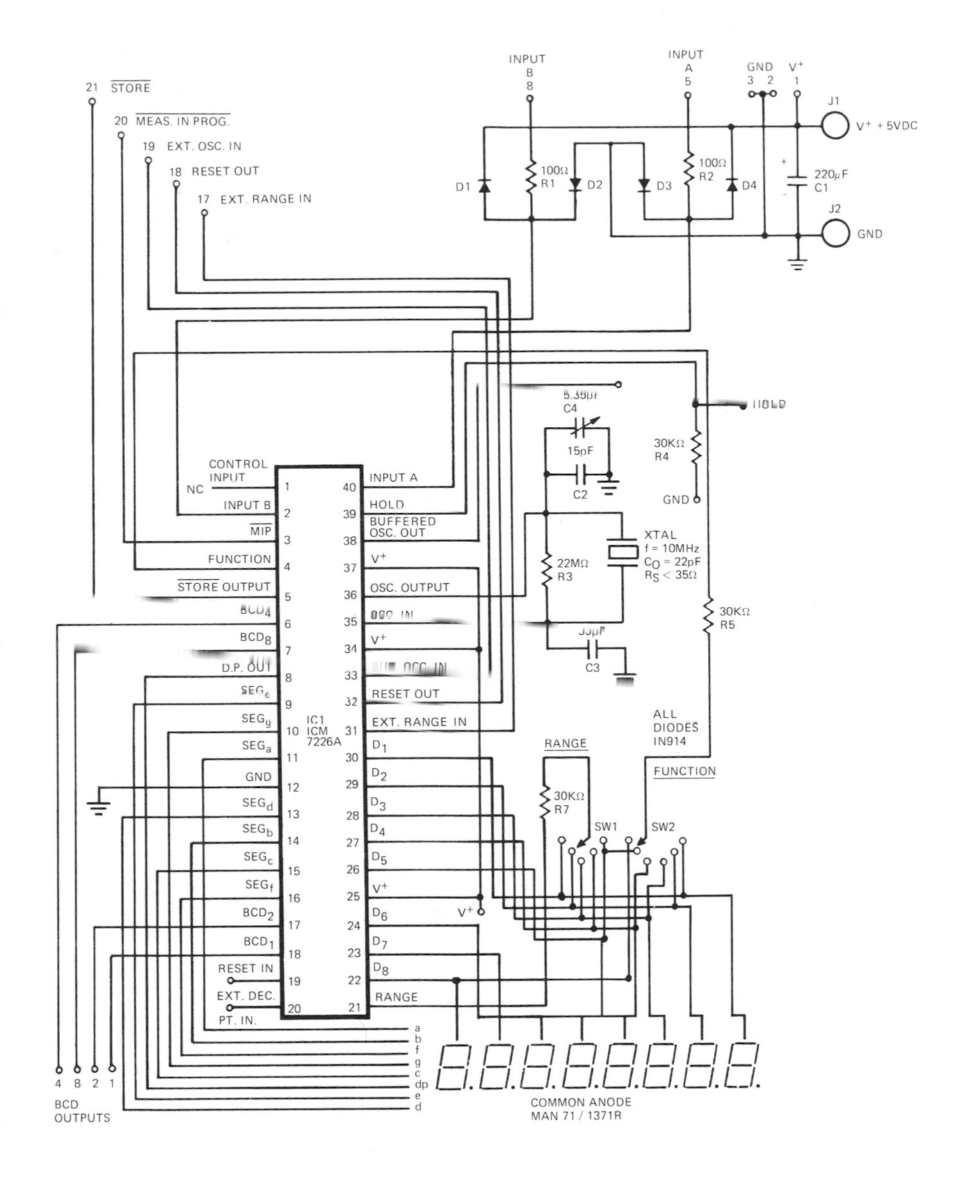

Figure 6-11. Intersil ICM7226A/B 10MHZ universal counter.

Intel 8254 Programmable Interval Timer

The Intel 8254, shown in Figure 6-12, is a counter/timer device designed to solve the common timing control problems in microcomputer system design. It provides three independent 16-bit counters, each capable of handling clock inputs of DC to 8 MHz (8254) and DC to 10 MHz (8254-2). It has six programmable counter modes, Status Read-Back Command, Binary or BCD counting, and operates from a single +5 VDC @140 mA power supply. It uses HMOS technology, comes in a 24-pin plastic or CERDIP package and is compatible with most microprocessors. For additional information contact Intel Corporation, 3065 Bowers Avenue, Santa Clara, CA 95051. Similar programmable counter/timer devices are made by Motorola (MC6840 Programmable Timer Module) and Zilog (Z80-CTC Counter/Timer Circuit).

The use of a programmable counter allows easy interfacing of a frequency or count to a computer. The computer first resets the counters, then gates on the counters. The TTL signal is fed into CLK 0. If more than 16-bits (65536 counts) are needed the OUT 0 is connected to CLK 1. This gives a total of 32 bits or a total count of 4,294,967,296. One counter is still left for another function.

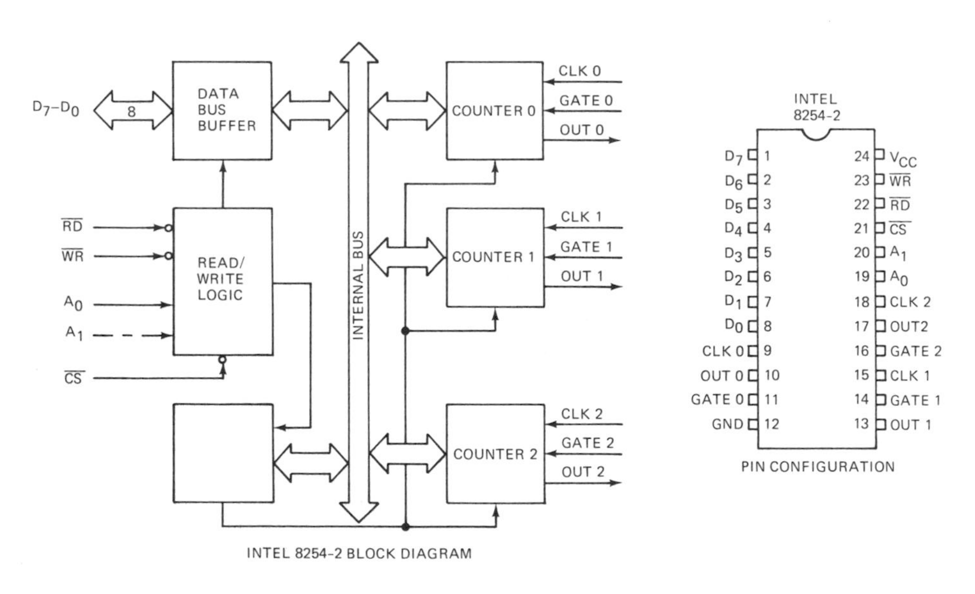

Figure 6-12. Intel 8254-2 programmable interval timer.

SECTION 2

HIGH SPEED PHOTOMETRY

7. OCCULTATIONS BY PLANETS AND SATELLITES

Robert L. Millis

I. INTRODUCTION

Our solar system is a scene of constant motion. Planets revolve around the Sun. Satellites traverse their orbits about the planets. Countless asteroids move across the sky, while an occasional comet sweeps in from the most distant regions of the solar system. The earthbound observer views this seeming maelstrom against the backdrop of distant stars. It is not surprising that frequently one of these moving bodies passes directly between the observer and a more distant object, momentarily blocking that object from view. Similarly, a satellite sometimes passes through the shadow cast by its primary, by another satellite, or by rings and, accordingly, is for a time eclipsed.

Careful observation of these naturally occurring experiments can result in valuable scientific discoveries. In 1675, Ole Roemer, by analyzing timings of eclipses of the Galilean satellites, demonstrated that the velocity of light is finite. Had it occurred to him, Roemer could have calculated a reasonably good value for that fundamental constant. By the middle of the eighteenth century, the observed abruptness of stellar occultations by the Moon was used effectively as evidence against the presence of a significant atmosphere on our nearest celestial neighbor. More recently, in 1977, astronomers monitoring the passage of Uranus in front of a nondescript, 9th magnitude star, discovered a remarkable system of nine narrow rings encircling that distant planet. These examples represent only a fraction of the important scientific results obtained from occultation studies. (For simplicity, I will define the term "occultation studies" to include phenomena which are actually eclipses. This distinction is one of viewpoint, because an observer on the eclipsed object would, in fact, be witnessing an occultation.)

While certain occultations and eclipses require access to large telescopes and sophisticated instrumentation for successful observation, many can be well observed with modest photoelectric equipment of the type increasingly found at small observatories. Useful work can be done with a telescope of 10- to 20-inches aperture and a simple single-channel photometer equipped with either DC or pulse-counting electronics. The only special requirement of the data recording system is that it be able to record continuously at a time resolution of a second or better for intervals up to an hour. Often a simple strip-chart recorder will suffice, although a digital device is to be preferred.

The pursuit of occultations has two attributes, in addition to high scientific potential and minimal instrumental requirements, which should make this work particularly attractive to students and amateur astronomers. First of all, observing occultations is exciting. An occultation or eclipse is a unique event. It happens at a particular time and is visible only within a limited--sometimes a very limited--area. Your efforts to observe the event can be defeated by weather, by equipment failure, by prediction error, or by your own blunders. As the predicted time draws closer, the suspense builds. If you miss the occultation, it is gone forever. If you are successful, the elation is tremendous.

The second somewhat unique attribute of occultation work is that the observer usually knows before leaving the telescope site whether he has been successful. These events often have short time scales and produce large signal changes. You see things happening in real time, and a useful result can be achieved in a single, brief observing session.

Two important types of occultations, occultations of stars by asteroids, and lunar occultations, are discussed in the following two chapters of this book. The present chapter will be confined to events involving other members of the solar family, namely, planets, satellites, rings, and comets. Some of the phenomena we will discuss occur frequently, while others are extremely rare. In each case the scientific value of observations will be described and suggested observational procedures spelled out. We will begin with the most easily observed phenomena and proceed to those which are observationally and instrumentally more challenging.

II. ECLIPSES OF THE GALILEAN SATELLITES

Even the most casual observer of the planets has noticed the ever-changing configuration of Jupiter's four large moons. Sometimes all four moons are seen arrayed to either side of the planet. At other times, one will be apparently missing, because it is either behind the planet, directly in front of the planet, or lost in Jupiter's shadow. Occasionally more than one satellite will be hidden. If the observer is lucky or if he has had the forethought to consult predictions in the *Astronomical Almanac* or the *Handbook of the British Astronomical Association*, he will occasionally see a satellite, standing well away from the planet's limb, gradually disappear into or emerge from Jupiter's shadow. Because the Galilean satellites are in orbits inclined by only about 3° relative to the orbital plane of Jupiter, these eclipses occur, with one exception, each time the moons circle the planet. The exception is Callisto, the outermost of the four, which passes above or below the shadow at times when one pole or the other of its orbit is tipped toward the Sun.

As was mentioned in the Introduction, eclipse disappearances and reappearances of the Galilean satellites have been observed almost from the time the telescope was invented. The period of most intensive observation was between 1878 and 1903, when E. C. Pickering and his colleagues at Harvard College Observatory used a series of ingenious visual photometers to observe 706 of these events (Pickering 1907). R. A. Sampson's theory describing the motion of the Galilean satellites (Sampson 1909) was based on these observations. Even today, the ephemerides of the satellites published in the Astronomical Almanac are computed from Sampson's tables.

To this writer's knowledge, the first photoelectric observations of ordinary eclipse phenomena of the Jovian satellites were made in 1950 by G. P. Kuiper (Harris 1961). The number of events observed photoelectrically during the next fifteen years can be counted on one's fingers. Between 1965 and 1980, observational interest in eclipses of the Galilean satellites increased, reaching its peak in the early 1970's. At present, few, if any, of these phenomena are being observed at visual wavelengths by professional astronomers.

The scientific objectives of the photoelectric observations made of these events during the past 30 years fall into three basic areas: (1) refinement of the ephemerides of the satellites, (2) probing the characteristics of the Jovian atmosphere, and (3) measurement of various properties of the satellites, including possible albedo changes during eclipse.

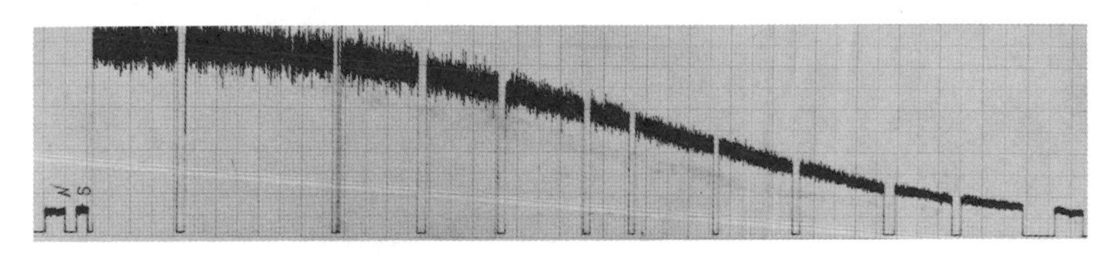

Figure 7-1. A lightcurve of the emergence of Callisto from Jupiter's shadow observed by the author with the Lowell Observatory's 21-inch telescope on November 27, 1974. The deflections labeled "N" and "S" are sky measurements north and south of the satellite following reappearance. The periodic interruptions of the signal were made to permit recentering of the satellite in the photometer's entrance aperture.

Figure 7-1 shows a typical strip-chart recording of an eclipse reappearance. The lightcurve of an eclipse disappearance would, of course, be essentially a mirror image of the one shown in Figure 7-1. As a satellite moves into eclipse, it first encounters the penumbral portion of Jupiter's shadow. The brightness of the satellite initially fades rather slowly and then more rapidly as the moon enters the umbra. The exact shape of this portion of the lightcurve is largely determined by the geometry of the situation and the diameter and albedo distribution of the satellite. Once inside the umbra, the satellite continues to be faintly illuminated for a while because of light refracted into the umbra by Jupiter's atmosphere. (Earth's moon is visible during total eclipse due to the same phenomenon.) The shape of this faint, so-called refractive tail on the lightcurve is controlled by the density structure of the Jovian atmosphere, the sources of opacity in the atmosphere above the cloud tops, and the altitude of the cloud deck. T. F. Greene, R. W. Shorthill, D. Smith, and their colleagues at the University of Washington and M. Price at the Planetary Science Institute have constructed elaborate mathematical models to compute theoretical lightcurves which could then be matched with observations to determine the various parameters discussed above (Greene, Smith, and Shorthill 1980; Price and Hall 1971). Observationally, this type of work is beyond the scope of the present book because it depends on accurately observing that portion of the lightcurve where the satellite's brightness has been diminished by a factor between 100 and 10,000! Obviously, very large telescopes are required.

In 1964, A. B. Binder and D. P. Cruikshank (1964) reported that they had observed the inner Galilean satellite, Io, to be anomalously bright upon reappearance from eclipse. The excess brightness reportedly persisted for approximately 10 to 15 minutes. These authors interpreted the brightening in terms of a frost or snow which formed while the satellite was in eclipse. When the satellite emerged again into sunlight, its surface warmed and the bright deposit gradually dissipated. Astronomers at several major observatories tried for fifteen years to establish whether post-eclipse brightening of Io, in fact, occurs. Although it would seem at first thought an easy problem, since Io is a 5th magnitude object, the difficulty arises in accurately subtracting the background of scattered light from Jupiter. Io, at best, reappears from eclipse at about one Jovian radius from the limb of the planet. As the satellite moves rapidly away from the limb after reappearance, the contribution of sky background to the total signal drops steeply. One must be able to accurately allow for this effect, or a spurious post-eclipse brightening may result (Franz and Millis 1974). About the only thing on which astronomers have been able to agree is that if the effect occurs at all, it is intermittent. Voyager scientists were unable to detect any brightening following the three reappearances of Io that they observed (Veverka et al. 1981). While investigation of post-eclipse brightening is certainly feasible with a small telescope and simple photometer, individuals contemplating such a study should carefully research the scattered light problem as described in the references at the end of this chapter. If the professionals cannot convince each other that they are doing the job correctly, they are unlikely to be easily persuaded by someone outside their circle.

Perhaps the best reason for observing eclipses of the Galilean satellites in the post-Voyager era is the astrometric information that such observations provide. The very fact that an eclipse disappearance or reappearance is occurring places tight constraints on the orbital longitude of the satellite involved. It is very easy using photometric observations like those in Figure 7-1 to determine the time of a particular phase of the eclipse--for example, the half-intensity point--to within an uncertainty of ±10 seconds. Assuming the eclipse can be modeled properly, this level of timing uncertainty translates to an uncertainty in the position of the satellite along its orbit on the order of ±100 km. To achieve the same positional accuracy by conventional astrometric methods would require that the uncertainty in the measured positions of the satellites be less than ±0.03 arcsec--a difficult task indeed.

Observation of ordinary eclipses of the Galilean satellites for purposes of timing is a relatively simple task; the events occur slowly, with a representative time scale for reappearances and disappearances of Io and Europa being 5 minutes, increasing to 20 minutes for Callisto. The figure for Ganymede falls between these extremes. Integration times of 5 to 10 seconds are appropriate, although the midtime of each integration should be known to better than 1 second. The only significant observational problem is coping with the steeply sloping background of scattered light, as we have discussed earlier. The problem is less serious for the outer satellites than the inner ones (for obvious reasons) and will be most serious for all satellites near opposition and least serious near quadrature. Disappearances are most easily observed before opposition, while reappearances are more favorable following opposition. The reader need only draw a diagram showing the relative positions of the Sun, Earth, Jupiter, and satellite at the times of the various configurations to convince himself of the correctness of the above statements. Consulting the figures accompanying the predictions for these events given in the Astronomical Almanac will also be helpful.

Several steps can be taken to minimize the scattered light problem, including keeping the telescope optics clean, using the smallest practical entrance aperture in the photometer, observing in a visual or red passband, and guiding accurately during the observation. Accurate guiding is facilitated by installing a pellicle in the photometer which reflects a few percent of the light into the aperture-viewing eyepiece while the remainder passes through to the photomultiplier. With this arrangement the satellite image can be maintained in the precise center of the entrance aperture without the need for periodically interrupting the observations. Having taken these precautions, it will still be necessary to measure the sky brightness at the position of the satellite. For disappearances, the job can be accomplished by measuring sky brightness one aperture diameter north and south of the satellite prior to the eclipse. The satellite is then centered in the aperture, and the disappearance recorded until the satellite has completely faded, at which time you are measuring the sky brightness at the satellite's final visible position. The task is more difficult for reappearances since it is usually not possible to set the entrance aperture on the precise spot where the satellite will reappear. Ordinarily, one must wait until the satellite can be seen before hastily centering it in the photometer aperture and recording the remainder of the event. At a minimum, the sky should be measured north and south of the satellite following emersion from the shadow; and, if the event is a long one, sky measurements should be made shortly after the object is first acquired, as well.

III. MUTUAL OCCULTATIONS AND ECLIPSES

In addition to being eclipsed regularly by their parent planet, the Galilean satellites sometimes eclipse and occult one another. These so-called mutual events occur every six years around the times of the Jovian equinoxes when the Sun and the Earth pass through the equatorial plane of Jupiter. Typically, in such a season, a few hundred occultations and eclipses occur spread over an interval of several months.

The first reported observation of a mutual event was by a Saxon farmer named Arnold, who in 1693 saw an occultation of Europa by Ganymede. Two centuries later, a Spanish astronomer, Comas Solá, was, quite by accident, the first to witness a mutual eclipse (Sandner 1965). Until the early 1970's, mutual phenomena remained almost exclusively the business of amateur astronomers, and even they showed little observational interest. Predictions of the times at which individual mutual events would occur have appeared sporadically since 1926 in amateur-oriented publications such as the _Handbook of the British Astronomical Association_ and _The Strolling Astronomer_. Occasional accounts of visual observations can be found in the popular literature, but if professional astronomers were even aware of the occurrences of the mutual events, they were strangely silent on the subject.

In scanning the literature prior to 1972, I have found only one reference to the mutual phenomena by a professional astronomer. George Hall Hamilton was observing at Lowell Observatory's 24-inch refractor on the night of December 5, 1919, when he noticed that two of the Galilean satellites were apparently approaching each other. Recalling that the Earth was near the equatorial plane of Jupiter at the time, Hamilton turned his attention to the satellites and was soon rewarded by an occultation of Io by Ganymede. Under what he described as nearly perfect seeing conditions, Hamilton produced a drawing which was published in _Popular Astronomy_ (Hamilton 1920). Realizing that mutual occultations and eclipses could be used to good advantage in studying the Galilean satellites, he urged that such observations be "pushed with all vigor." However, more than a half century passed before his advice was heeded.

Early in 1972, R. T. Brinkmann at the Lunar Science Institute computed predictions for a series of more than 300 mutual occultations and eclipses which would occur between February 1973 and May 1974. In a widely circulated preprint, Brinkmann described the upcoming opportunity and discussed in detail the scientific potential of mutual event observations in three specific areas: (1) refinement of satellite ephemerides, (2) determination of satellite diameters, and (3) mapping of albedo features on the surfaces of the satellites. Stimulated by Brinkmann's paper, an international campaign was organized, involving both professional and amateur astronomers, to record photoelectrically as many of the 1973/1974 mutual events as possible (Brinkmann and Millis 1973). Of 174 potentially observable occultations and eclipses, at least 84 were successfully observed at one or more sites. Lightcurves of three of these events are shown in Figure 7-2.

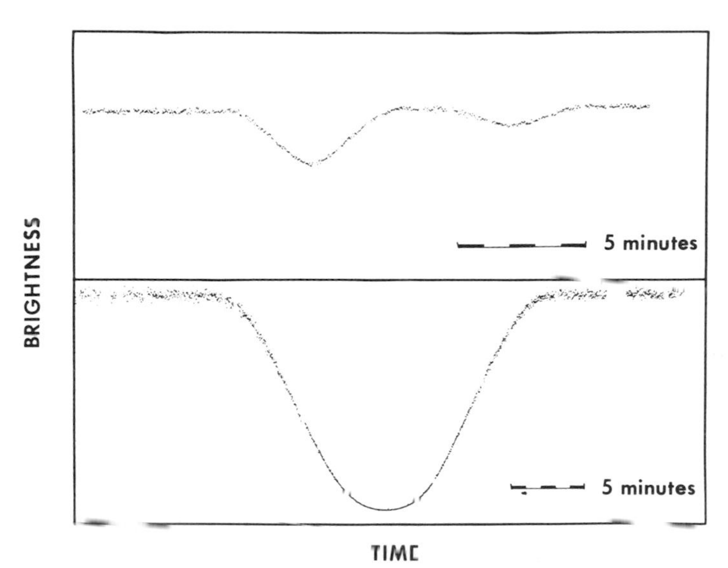

Figure 7-2. Mutual event lightcurves. The top curve shows a partial occultation of Europa by Io, followed almost immediately by a partial eclipse involving the same two satellites. The observations were made by Peter Birch at Perth Observatory on August 2, 1973. The bottom curve is a lightcurve of a nearly total eclipse of Europa by Ganymede observed by the author at Lowell Observatory on September 17, 1973. This figure is reproduced courtesy of the editor of Mercury (Millis 1974).

That mutual event observations are ideal for ephemeris refinement is easy to understand, since these phenomena occur only when the two satellites involved and the Sun or Earth achieve a precisely defined configuration. Using photoelectric lightcurves like those in Figure 7-2, the midtimes of the individual events can be determined to within a few seconds. A large number of such timings places very tight constraints on the satellite ephemerides

The shape of a mutual event lightcurve, as distinct from its timing, is primarily controlled by the diameters of the satellites involved, the surface markings of the eclipsed or occulted satellite, and the impact parameter (i.e., the degree to which the occultation or eclipse departs from being central). Hence, by modeling the lightcurves one could expect to derive better values for the diameters of the three satellites for which this parameter was not already well known. There was also reason at least to hope that some information about albedo features could also be obtained.

By far the most complete analysis of the 1973/1974 mutual event data was performed by K. Aksnes and F. A. Franklin (1976) at the Center for Astrophysics. Using 91 photoelectric lightcurves, they were able to derive values for the diameters of Europa, Ganymede, and Callisto which agree well with subsequent Voyager results. The diameter of Io was already accurately known from stellar occultation observations. These investigators also improved the latitude and longitude residuals of Sampson's revised theory of motion of the Galilean satellites by as much as a factor of 10. They were unable, however, to derive any definitive information about the distribution of light and dark material on the satellite surfaces.

The 1973/1974 mutual event season was especially favorable. In 1979, on the other hand, these events occurred around the time of Jupiter's conjunction, and relatively few were observed. We can look forward to a much better opportunity in 1985/1986, when the events will be more nearly centered on opposition. Although the diameters and albedo distributions of all four Galilean satellites are now well known as a result of the Voyager encounters, observation of mutual events remains an extremely powerful tool for ephemeris refinement. It is probable that the major observatories will be far less active in observing these events in 1985/1986 than they were in 1973/1974. Consequently, adequate observational coverage will depend on individuals at smaller observatories.

The observational requirements and procedures for mutual events are very similar to those described earlier for ordinary eclipses of these satellites. The durations of mutual events are typically measured in minutes to tens of minutes, so integration times on the order of ten seconds are appropriate. As before, the midtime of each integration should be known to better than one second. On average, scattered light from Jupiter will be less of a problem for mutual events than for ordinary eclipse reappearances and disappearances, since the former can occur significantly farther from Jupiter's limb. Nevertheless, sky measurements should be made on at least two sides of the satellites involved, both before and after each event. If possible, it is also extremely important to measure the individual brightness of each satellite involved in an occultation while the two are well separated, either before or after the event (or both). The same should be done for eclipses if the two objects involved are so close together during the eclipse that both must be included in the aperture of the photometer. A V filter is recommended for these observations, to facilitate comparison with measurements made elsewhere.

Mutual events are not restricted to Jupiter's satellites. Certain satellites of Saturn and the five moons of Uranus also produce these phenomena. In fact, a few events involving Saturn's satellites were actually recorded photoelectrically for the first time during 1979/1980. However, because of the faintness and angular proximity to the planet of both sets of satellites, observation of these events is beyond the reach of this book's intended audience. The reader can console himself with the realization that the next opportunity to observe Saturn satellite mutual events is more than a decade away, while such happenings for the Uranian satellites will not take place until well after the turn of the century.

IV. PLUTO-CHARON PHENOMENA

In 1978, J. W. Christy of the U. S. Naval Observatory discovered a faint satellite orbiting the planet Pluto (Christy and Harrington 1978). This difficult object, which has been named Charon, circles Pluto with a period of 6.3871 days. Since Pluto's rotational lightcurve has the same period, it is believed that the two objects are in the unique situation of being tidally locked, perpetually keeping the same sides facing each other.

As seen from Earth, Charon is never more than one arcsecond from Pluto. Consequently, precise specification of the satellite's orbital parameters and physical properties has not been possible. Charon's diameter is believed to be between one-third and one-half that of Pluto, but the size of that distant planet is, itself, poorly known. A single stellar occultation observation by A. R. Walker (1980) has put a lower limit of 1200 km on the diameter of the satellite.

Within the next few years, photometrists will have a valuable opportunity to significantly extend our knowledge of the Pluto-Charon system. As was first noted by L. E. Andersson (1978), the orbit of Charon is now approaching an edge-on orientation. Consequently, a long series of eclipses, occultations, and transits spanning five or six apparitions will soon commence. The most recent work by R. S. Harrington and J. W. Christy at the U. S. Naval Observatory and by E. K. Hege _et al_. using the Multiple Mirror Telescope indicates that the series will begin sometime in 1983, 1984, or 1985 (Harrington and Christy 1981; Hege _et al_. 1982). Central events should last on the order of five hours and will produce changes in the combined brightness of the two bodies near 0.2 mag. Accurate lightcurves of several such events will permit the diameters and albedos of Pluto and Charon to be measured precisely. These observations will also lead to major improvement in Charon's orbital elements.

The Pluto-Charon phenomena present a distinct observational challenge to observers with small- to moderate-sized telescopes. During the interval of interest, Pluto will be near 14 mag in V. Consequently, simply finding the object will be a nontrivial task for some. A dark sky and an accurate finding chart, such as the one published annually in the *Handbook of the British Astronomical Association*, will be of significant help in the process. Having found Pluto, one is then faced with the problem of measuring its brightness with an uncertainty of not more than one or two percent. With a 20-inch telescope, a V filter, and typical values for extinction, reflectivity of optics, and quantum efficiency, one would expect on the order of 100 counts per second from Pluto and Charon combined. At a good site, the contribution from the sky background will be comparable if a reasonably small entrance aperture is used. Under these circumstances, the integration time on object and sky needed to reach the desired accuracy will be measured in minutes rather than seconds. The exact value depends critically on the individual observer's equipment and on sky quality.

The recommended procedure for observing the Pluto-Charon events is identical to that used by many variable star observers. Two nearby comparison stars whose brightnesses and colors are similar to Pluto's are selected. The brightness of the first star is measured, then that of Pluto, then that of the other comparison star. This cycle is repeated throughout the observing interval. One then reduces the data to magnitude differences between Pluto and one of the comparison stars and magnitude differences between the two comparison stars as a function of time. Reducing the data in this way minimizes the effects of errors in extinction coefficients, drifts in instrumental sensitivity, etc. Furthermore, the degree of stability observed in the magnitude difference between the two comparison stars provides a good guide to assessing the reliability of any variations observed in the brightness of Pluto.

As has been mentioned, the times of individual Pluto-Charon phenomena cannot, as yet, be predicted accurately. However, it is obvious that these events will occur near Charon's conjunctions. The approximate time of conjunction can be derived from the table of times of greatest northern elongation given in the *Astronomical Almanac*. Once a few events have been observed, more accurate predictions will undoubtedly be published.

V. STELLAR OCCULTATIONS

So far in this chapter, attention has been focused on occultations and eclipses of one solar system member by another. Even more powerful events in terms of scientific potential are those in which the occulted body is a distant star. Stellar occultations permit measurement of dimensions of solar system objects with an accuracy unapproached by other ground-based techniques. Furthermore, the observations will reveal the faintest wisp of an atmosphere on the occulting body and perhaps will yield temperature and other atmospheric characteristics. Stellar occultations provide valuable opportunities to measure the amount of light transmitted as a function of position in planetary rings and in the comae of comets. Finally, it is sometimes possible to measure the angular diameter of the occulted star from high-quality occultation data.

Stellar occultations by the Moon and by minor planets are discussed elsewhere in this book. These types of events occur frequently and are readily observable with small telescopes. Therefore, they should be of particular interest to readers of this book. Discussion in the present chapter will be confined to stellar occultations by other solar system bodies, namely planets, satellites, rings, and comets.

Stellar Occultations by Planets and Satellites

The first photoelectric observations of an occultation of a star by a planet were made in 1952 by W. A. Baum and A. D. Code (1953), who recorded the occultation of σ Arietis by Jupiter. The May 14, 1971 occultation of β Scorpii C by Io was the first satellite event to be observed (O'Leary and Van Flandern 1972). Since these beginnings, occultations by all of the planets except Mercury, Saturn, and Pluto have been observed from terrestrial observatories, as have events involving the satellites Ganymede and Charon (see Elliot 1979).

It should be noted at the outset that opportunities to observe these types of events with conventional, blue-sensitive photometric equipment are relatively rare. Although astronomers have devised various schemes for reducing the light contribution from the planet, the closer planets still require naked-eye stars to produce an observable event. Estimates by B. O'Leary (1972) indicate that such events will be separated by intervals of decades, if not centuries. In addition, pinpointing the region of visibility of an occultation by Pluto or most outer planet satellites is at present beyond our capabilities. As a result, the observer at a small, fixed observatory may not have the opportunity in his lifetime even to witness, not to mention adequately observe, one of these events.

The character of the lightcurve of a stellar occultation by a planet or satellite and the type of information that can be derived from the data are determined by whether or not the occulting body possesses an atmosphere. If it does not, the occultation is equivalent to a lunar occultation or to an event involving a minor planet. With adequate observational coverage, one can expect to determine the size and shape of the occulting body and perhaps the angular diameter of the occulted star. The reader is referred to the chapters by Harris and Blow for observational and analytic details.

If the occulting body has an atmosphere, as do Titan and all the planets except Mercury, the disappearance and reappearance of the star will be much more gradual than for an occultation by an airless body. The reason is illustrated in Figure 7-3. At the beginning of the occultation, the light rays from the star first encounter the upper portion of the occulting body's atmosphere. The rays are bent by refraction in the atmosphere, as shown in the figure. As immersion progresses, the light rays penetrate to lower and lower levels in the atmosphere where the increasing atmospheric density will cause progressively greater refractive bending. Consequently, the light from the star is increasingly fanned out so that as time passes fewer and fewer photons fall within the fixed collecting area of the observer's telescope. Long before the star light reaches a depth in the atmosphere where significant atmospheric extinction occurs, the differential refraction will have caused the star to have disappeared from view.

Baum and Code (1953) showed that if the occulting atmosphere were the same temperature throughout, the immersion and emersion lightcurves would be smooth and could be represented by a simple equation involving atmospheric temperature and mean molecular weight as the only unknown parameters. In fact, the actual conditions in planetary atmospheres have turned out to be more complex. Figure 7-4 shows the emersion lightcurve for the April 22, 1982, stellar occultation by Uranus. Note that the lightcurve is not smooth but displays many abrupt, positive-going spikes. It is generally agreed that these brief flashes are caused by density fluctuations in the occulting atmosphere. The nature of the density fluctuations and the appropriate methods of analyzing the lightcurves are, however, the subject of considerable disagreement. The nitty-gritty of the controversy can be found in the recent review by Elliot (1979). We will be content here to simply note that the method based on numerical inversion of the lightcurve developed by Cornell University astronomers gave values of temperature, pressure, and number density for the upper region of Mars' atmosphere in good agreement with subsequent *in situ* measurements by the Viking landers (Elliot *et al*. 1977).

Observations of occultations by bodies with atmospheres can also be used to derive the object's size and shape, just as has been done for asteroids and satellites. The only difference is that the resulting dimensions will pertain not to the solid surface but to a particular number density level in the atmosphere.

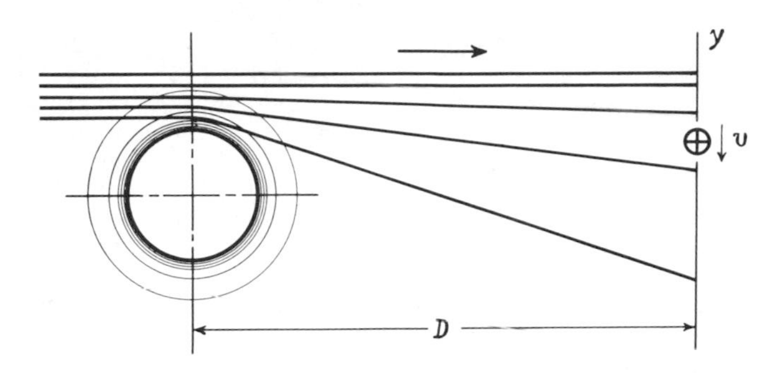

Figure 7-3. A diagram showing the differential refraction of star light in a planetary atmosphere during a stellar occultation. Parallel rays of light from the star enter the diagram from the left. The rays are bent by refraction as they pass through the planet's atmosphere, with those penetrating most deeply in the atmosphere being bent the most. The observer on Earth at a distance, D, can be imagined to move across the pattern of light rays with a velocity, v . As the rays become increasingly fanned out, the observer's telescope collects less and less light from the star. This figure was reproduced from the Astronomical Journal by permission of the editor (see Baum and Code 1953).

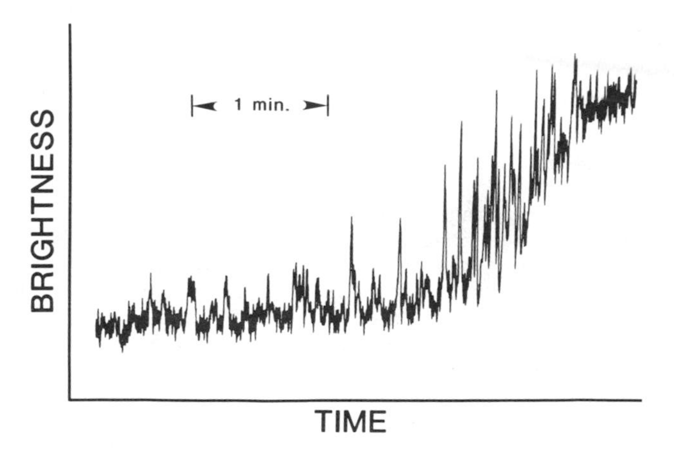

Figure 7-4. The emersion lightcurve of the April 22, 1982 stellar occultation by Uranus. The observations were made by the author using the 1.5-meter Infrared Flux Collector at the Cabézon Observatory in the Canary Islands. The spikes in the lightcurve result from density fluctuations in Uranus' atmosphere.

The best technique in observing stellar occultations by planets is to minimize the contribution of light from the planet while maintaining enough signal from the star for a good signal-to-noise ratio. For the nearby planets, one may often use a small enough entrance aperture that only a portion of the planet's limb is included. For events involving more distant planets such as Uranus, Neptune, and Pluto, the entire planet and the star must be included in the photometer's aperture. One must then search for a bandpass where the spectrum of the planet is depressed relative to the star's. Very good success has been achieved in recent years for occultations of relatively faint stars by Uranus and Neptune, by observing in the deep methane absorption bands present in the near-infrared portions of these planets' spectra (Elliot, Veverka, and Millis 1977). Unfortunately, relatively expensive interference filters and red-sensitive photomultipliers are required.

It is desirable in observing occultations by planets to record data continuously at a time resolution on the order of 0.01 sec commencing many minutes before immersion and continuing until a like interval after emersion. This procedure allows for errors in the predicted times of immersion and emersion, for the discovery of previously unknown rings and satellites, and for recording the central flash--a refractive focusing of the star light seen near mid-event along the center of the occultation track. Accurate guiding is important and can be accomplished by the use of a pellicle as described earlier in this chapter.

Stellar Occultations by Planetary Rings

Perhaps the best-known result of stellar occultation observations has been the discovery of the rings of Uranus. On March 10, 1977, astronomers attempting to observe the occultation of SAO 158687 by Uranus recorded several brief dips in signal while the star was well outside the planet's limb (Elliot, Dunham, and Mink 1977; Millis, Wasserman, and Birch 1977). Examples of these unexpected features in the lightcurve are shown in Figure 7-5. It was quickly realized that the dips resulted from occultations of the SAO star by nine narrow rings in Uranus' equatorial plane. The radii, eccentricities, precession rates, widths, and optical thicknesses of the rings have been accurately derived from the 1977 observations and those obtained during several subsequent occultations (see Nicholson, Matthews, and Goldreich 1982). These accomplishments are truly remarkable in view of the fact that the rings cannot be directly resolved by the most powerful earthbased telescopes and have only barely been detected in reflected sunlight.

Earthbased observers have had less success in observing occultations by other planetary ring systems. Because the view from Earth is always of the sunlit side of Saturn's rings, a very bright star is required to be visible against the bright ring foreground. Such events have been observed visually, but photoelectric observations from the ground have not yet been successful. The photopolarimetry experiment on the Voyager II spacecraft, however, did obtain excellent observations of a stellar occultation seen through that portion of Saturn's rings that was shadowed by the ball of the planet (Lane *et al*. 1982). Jupiter's ring apparently contains little material. Occultation observations from the Kuiper Airborne Observatory (Dunham *et al*. 1982) failed to reveal any diminution of the star light. In the case of Neptune, occultation searches for rings are still inconclusive. Observations of two occultations in 1981 gave no evidence of rings (Elliot *et al*. 1981), although University of Arizona astronomers apparently did detect a new satellite (Reitsena *et al*. 1981). These observations did not, however, sample the entire region surrounding Neptune where rings might be present. Very recently, observations of a 1968 occultation by Neptune have emerged which seem to indicate the presence of a ring system relatively near the planet (Guinan, Harris, and Maloney 1982). Additional occultation observations will be required to confirm this result.

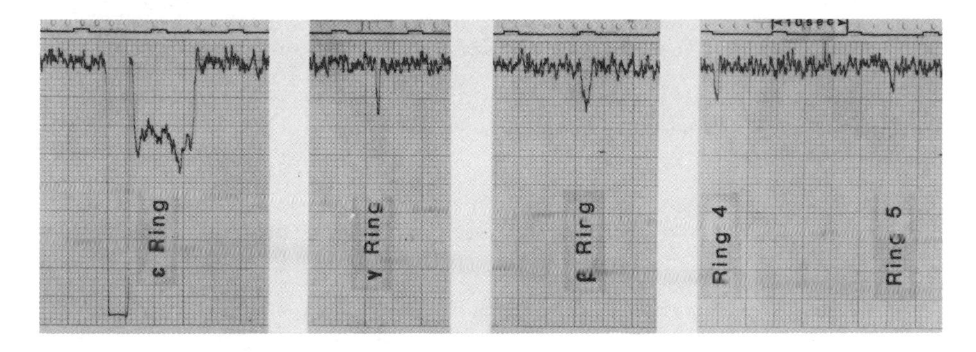

Figure 7-5. Strip chart showing the occultations of SAO 158687 by various Uranian rings observed on March 10, 1977, by the author at Perth Observatory. These observations and similar ones of the March 10 event from other sites resulted in the discovery of the rings.

Observationally, occultations by the rings of Uranus and Neptune (if they exist) are equivalent to occultations by the planets themselves. The same bandpasses, integration times, and observational procedures discussed in the previous section are appropriate.

Stellar Occultations by Comets

Comets are extended objects which move rapidly across the sky. As such, they must frequently occult relatively bright stars. Most of these occultations are scientifically uninteresting because in the comet's tail and most of the coma there is insufficient material present to affect the star light in any detectable way. On occasion, however, the very central regions of the coma will pass in front of a star. In such instances, one might expect a measurable dimming of the star to occur. Precise photometric observations of one of these central occultations would make possible, among other things, a determination of the amount of dust as a function of radial position in the coma. Because cometary nuclei are believed to be only a few kilometers in diameter, there is little hope of detecting the nucleus itself by occultation techniques.

Recently, efforts have commenced at Lowell Observatory to predict stellar occultations by comets. Initial results have been encouraging, and it is hoped that predictions can soon be published. Comet Halley will be an obvious target of these efforts.

The observing procedure appropriate for a comet event is largely dependent on the surface brightness of the coma. As with all stellar occultations, it is desirable to choose a passband which depresses the contribution of light from the occulting body relative to that from the star. For comets, this will often mean avoiding the strong CN and C_2 emission bands near 3880 Å and 5100 Å, respectively. If the comet is relatively faint and compact, use a large entrance aperture in the photometer so that both the coma and the star will be included throughout the occultation. With a comparatively extended and bright comet like Halley, one would choose a small aperture centered on the star. In this case it will be necessary to determine the contribution to the total signal by the coma throughout the event. Probably the best way to make the measurement with a simple photometer is to offset the aperture along a line passing through the star and parallel to the comet's path either well before or after the occultation. Then, record the coma brightness as the comet moves past. Unless comets have a structure very different than we think, integration times as long as one second will be adequate.

ACKNOWLEDGMENTS

The author's occultation research is supported by NASA grant NSG-7603. Special thanks are due to O. G. Franz and L. H. Wasserman for their helpful critiques of this chapter.

REFERENCES

Aksnes, K., and Franklin, F. A. (1976). Mutual phenomena of the Galilean satellites in 1973. III. Final results from 91 lightcurves. Astronomical Journal 81, 464.

Andersson, L. E. (1978). Eclipse phenomena of Pluto and its satellite. Bulletin of the American Astronomical Society 10, 586.

Baum, W. A., and Code, A. D. (1953). A photometric observation of the occultation of Areitis by Jupiter. Astronomical Journal 58, 108.

Binder, A. B., and Cruikshank, D. P. (1964). Evidence for an atmosphere on Io. Icarus 3, 299.

Brinkmann, R. T., and Millis, R. L. (1973). Mutual phenomena of Jupiter's satellites in 1973-74. Sky and Telescope 45, 93.

Christy, J. W., and Harrington, R. S. (1978). The satellite of Pluto. Astronomical Journal 83, 1005.

Dunham, E. W., Elliot, J. L., Mink, D. J., and Klemola, A. R. (1982). Limit on possible narrow rings around Jupiter. Astronomical Journal 87, 1423.

Elliot, J. L. (1979). Stellar occultation studies. In Annual Reviews of Astronomy and Astrophysics, edited by G. Burbidge (Annual Reviews, Inc., Palo Alto), p. 334.

________, Dunham, E., and Mink, D. (1977). The rings of Uranus. Nature 267, 328.

________, French. R. G., Dunham, E., Gierasch, P. J., Veverka, J., Church, C., and Sagan, C. (1977). Occultation of Geminorum by Mars. II. The structure and extinction of the Martian upper atmosphere. Astrophysical Journal 217, 661.

________, Mink, D. J., Elias, J. H., Baron, R. L., Dunham, E., Pingree, J. E., French, R. G., Liller, W., Nicholson, P. D., Jones, T. J., and Franz, O. G. (1981). No evidence of rings around Neptune. Nature 294, 526.

________, Veverka, J., and Millis, R. L. (1977). Uranus occults SAO 158687. Nature 265, 609.

Franz, O. G., and Millis, R. L. (1974). A search for posteclipse brightening of Io in 1973. II. Icarus 23, 431.

Greene, T. F., Smith, D. W., and Shorthill, R. W. (1980). Galilean satellite eclipse studies. Icarus 44, 102.

Guinan, E. F., Harris, C. C., and Maloney, F. P. (1982). Evidence for a ring system of Neptune. Bulletin of the American Astronomical Society 14, 658.

Hamilton, G. H. (1920). Occultation of Satellite I by Satellite III in the Jovian system. Popular Astronomy 28, 141.

Harrington, R. S., and Christy, J. W. (1981). The satellite of Pluto. III. Astronomical Journal 85, 442.

Harris, D. L. (1961). In Planets and Satellites, edited by G. P. Kuiper and B. M. Middlehurst (The University of Chicago Press, Chicago), p. 327.

Hege, E. K., Hubbard, E. N., Drummond, J. D., Strittmatter, P. A., Worden, S. P., and Lauer, T. (1982). Speckle interferometric observations of Pluto and Charon. Icarus 50, 72.

Lane, A. A., Hord, C. W., West, R. A., Esposito, L. W., Coffeen, D. L., Sato, M., Simmons, K. E., Pomphrey, R. B., and Morris, R. B. (1982). Photopolarimetry from Voyager 2: Preliminary results on Saturn, Titan, and the rings. Science 215, 537.

Millis, R. L. (1974). The Galilean satellites. Mercury 3, 3.

________, Wasserman, L. H., and Birch, P. (1977). Detection of rings around Uranus. Nature 267, 330.

Nicholson, P. D., Matthews, K., and Goldreich, P. (1982). Radial widths, optical depths, and eccentricities of the Uranian rings. Astronomical Journal 87, 433.

O'Leary, B. (1972). Frequencies of occultations of stars by planets, satellites, and asteroids. Science 175, 1108.

________, and Van Flandern, T. C. (1972). Io's triaxial figure. Icarus 17, 209.

Pickering, E. C. (1907). Eclipses of Jupiter's satellites 1878-1903. Annals of the Astronomical Observatory of Harvard College 52, 1.

Price, M. J., and Hall, J. S. (1971). The physical properties of the Jovian atmosphere inferred from eclipses of the Galilean satellites. I. Preliminary Results. Icarus 14, 3.

Reitsema, H. J., Hubbard, W. B., Lebofsky, L. A., and Tholen, D. J. (1981). Discovery of a probable third satellite of Neptune. Bulletin of the American Astronomical Society 13, 721.

Sampson, R. A. (1909). A discussion of the eclipses of Jupiter's satellites 1878-1903. Annals of the Astronomical Observatory of Harvard College 52, 153.

Sandner, W. (1965). In Satellites of the Solar System (American Elsevier Publishing Company, New York), p. 41.

Veverka, J., Simonelli, D., Thomas, P., Morrison, D., and Johnson, T. V. (1981). Voyager search for posteclipse brightening on Io. Icarus 47, 60.

Walker, A. R. (1980). An occultation by Charon. Monthly Notices of the Royal Astronomical Society 192, 47P.

8. ASTEROID OCCULTATIONS

Alan W. Harris

I. WHY OBSERVE OCCULTATIONS?

Before discussing the scientific reasons for observing asteroid occultations, I wish to point out perhaps the most compelling reason of all - they are fun to observe! They provide a reason to travel to out of the way places that one would probably not otherwise visit, there is the rush of last minute updates in track position and the suspense of whether or not you will get an event, and finally the exchange of stories and data between other observers to find out where the track really went and if any secondary events were seen. Few observing programs of any real scientific value are as free of tedium but at the same time as challenging as observing asteroid occultations.

On the other hand, one would be reluctant to pursue an observing program for purely frivolous reasons. Successful observations of asteroid occultations are scientifically very valuable, as indicated by the substantial effort at present on the part of professional astronomers to observe these events. A paper by Millis and Elliot (1979) provides a thorough and very readable account of the scientific values and some of the pitfalls of occultation observing.

The surest scientific payoff of observations of an occultation is a very direct and unambiguous measurement of the asteroid's size and shape. By observing an event from many different locations, one can trace out a profile of the asteroid "shadow" on the earth, or equivalently, the profile against the sky (Figures 8-1 and 8-2). The other methods of diameter determination, combining absolute magnitude with albedo determined by infrared radiometry or polarimetry, each require some auxiliary assumptions about the nature of asteroid surfaces in order to compute a diameter from the observations, and hence, the results are to some degree dependent on the validity of these assumptions. These methods have been used to determine several hundred asteroid diameters. The occultation technique is free of any significant model errors. Given a sufficient number of observations, the profile of the asteroid is determined uniquely, and independent of any assumed property of the asteroid (other than that it is opaque!). Although it is impractical to observe hundreds of asteroid occultations adequately, it is possible to observe enough (several of each taxonomic class and spanning a

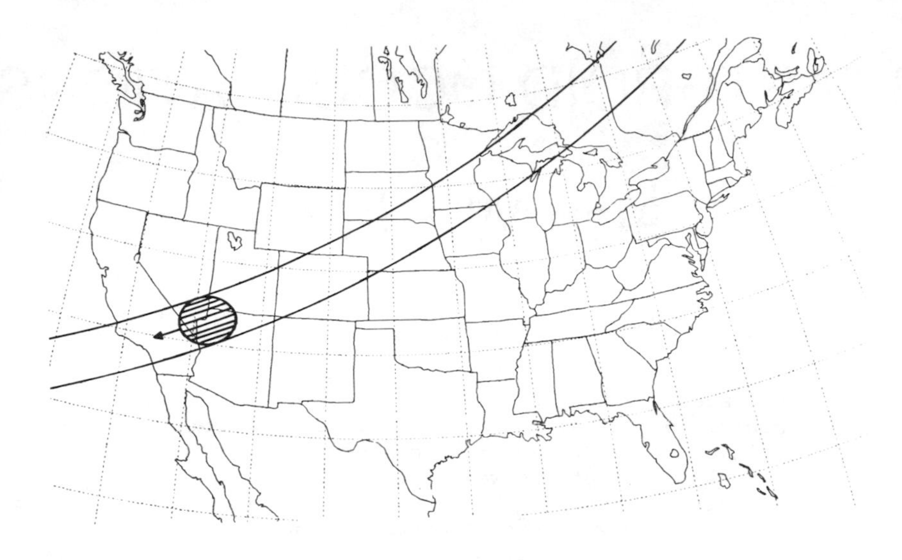

Figure 8-1. The predicted track of the occultation of the Star AG+0°1022 by the asteroid Juno on December 11, 1979. The "shadow" of Juno moved from Wisconsin to California in about 10 minutes.

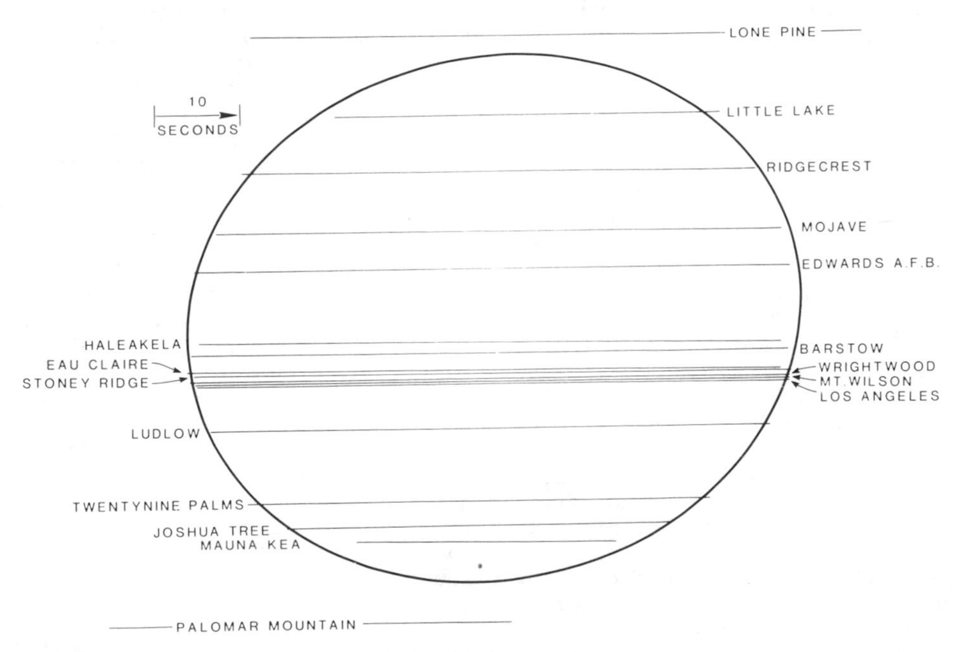

Figure 8-2. The profile of Juno from the December 11, 1979 occultation reconstructed by projecting the observed occultation chords back into the plane of the sky. This elliptical solution ignores small-scale irregularities in the limb which are apparent from some of the closely spaced chords.

significant range of size) to "calibrate" the indirect methods of diameter determination. The systematic "model" uncertainties in the radiometric and polarimetric diameter determinations have been estimated as $\sim$10%, whereas the precision of the measurements is no worse than a few percent. Hence, the goal of observations of an occultation is to determine the cross-sectional area of the asteroid to within a few percent. A less precise result is of little value, since it would not improve upon the radiometric and/or polarimetric determinations.

A second possible payoff of occultation observations, less certain but more dramatic, is the possibility of observing an occultation by a satellite of an asteroid. Such an event, if adequately verified, would be a headline discovery, certainly one of the most significant astronomical discoveries of the decade. If for no other reason that that, absolute verification is of paramount importance. What I wish to emphasize is that attempts to observe secondary occultation events are important and should be a part of the observing strategy. However, because of the need for absolute verification of an event, reliability and redundancy are more important than for observations of primary occultations. The evidence for binary asteroids is reviewed by van Flandern et al. (1979), and Millis and Elliot (1979) discuss the matter of observations of secondary events.

A third value of asteroid occultation observations is to study the star which is occulted. A significant fraction of occulted stars are discovered to be previously undetected binaries--too close for ordinary optical separation and yet not spectroscopically detectable for various reasons. This frequent outcome of occultation observations unfortunately confuses the question of possible secondary events, however the two situations are in principle unambiguous, and may, in fact, yield two chords per observing station as a bonus. It is also possible to measure the star's diameter by observing the diffraction pattern of the light just before and after the event. The time resolution required to study this phenomenon ($\sim$0.1 second) is generally less than for occultations by the Moon. In spite of this, measurements of star diameters by asteroid occultations must be regarded as relatively unimportant, since events are so rare compared to lunar occultations.

II. INSTRUMENTAL REQUIREMENTS

For the present and foreseeable future, it is impossible to predict occultations of asteroids smaller than $\sim$100 km in diameter to a useful accuracy (see Section III). Since the typical speed of an asteroid relative to the Earth is $\sim$10 km/sec, the duration of an average "central" occultation is $\sim$10 seconds or longer. We desire to obtain results with a precision of $\sim$1%, so the required accuracy of timing events is about 0.1 second. The best way to achieve this is to record WWV time signals simultaneously with the photometer output. If the photometer data recorder has its own internal clock, this can by synchronized to WWV to the required accuracy by ear, however do not count on being able to reliably mark the time of an event by observing the output of the photometer while listening to time signals. It is essential to record both time and the photometer output in some replayable form. It is not essential that the record be easily replayed for measurement, since the need to do so is infrequent. For example, one could record the photometer output as an audio tone on one channel of a cassette recorder while recording WWV on the other channel. Such a record could be measured quite precisely, although not necessarily easily.

At the present time, most predicted occultations involve stars brighter than ~10th magnitude. Hence, one need only have a system sensitive enough to "detect" a 10th magnitude star in 0.1 second. As a practical example, the chord identified as "Mohave" in Figure 8-2 was recorded (by an amateur) using an 8" telescope and a 1P21 photomultiplier. This event involved a star of visual magnitude ~9.0 occulted by Juno which was then visual magnitude ~8.5. The record quality was such that the 0.1 second timing accuracy was exceeded several times over, indicating that that system could achieve the desired accuracy for an occultation of any star brighter than ~10th magnitude. Eventually, occultation predictions may be extended to much fainter stars, so that one might desire to detect an occultation of a 12th magnitude star by a 10th magnitude asteroid. Such capability would require a telescope of ~16-inch aperture even with photon counting sensitivity. However, for the present, a modest system attached to a telescope of 8- to 10-inches aperture is adequate.

Since we are dealing with high signal levels and very short time intervals, there is little to be gained by cooling the photomultiplier tube. Also, since we are observing abrupt changes in light level, system stability is relatively unimportant; a sensitive DC amplifier is as good as a pulse-counting system for occultations.

Mechanical stability is of much greater importance. An abrupt error in telescope position, whether caused by wind, clockwork, touching the telescope, or whatever, can cause a loss of signal at a critical moment, or even worse, introduce spurious "events" so that the rest of the record cannot be trusted. It is essential that the telescope/photometer system be capable of producing a record of ~10 minutes duration without unaccounted losses in signal. The situation is somewhat eased by the fact that a fair amount of background skylight can be tolerated and therefore a large diaphragm aperture is acceptable. Under dark sky conditions, an aperture as large as several arcminutes in diameter can be used. However, a significant amount of moonlight may reduce the largest practical aperture to under one arcminute. A good design level to aim for is the capability to keep a star within a 30 arcsecond aperture for 10 minutes. For most small telescopes, this will require active guiding. I recommend a beamsplitter arrangement so that the actual beam entering the photometer is also examined visually. This aids in identifying spurious errors in the record and also allows the observer to actually see the occultation happen if the magnitude drop is large enough.

A final instrumental requirement is portability. A typical occultation track may be only 100-200 km across, hence in order to obtain a well-spaced grid of observations, site locations must be arranged to within ~10 km. This clearly requires portable units, however the locations are rarely so remote that electrical power cannot be had within an acceptable range of the desired site location. Since the darkest sky is generally not alongside a gas station, a portable power supply is desirable, but not strictly necessary.

One useful program can be carried out with a fixed site observatory. That is the search for occultations by satellites of asteroids. Since reliability of observation is a paramount requirement, the comfort and stability of a permanent site is an obvious advantage. The reason that portability is unnecessary is that the plausible range within which a satellite might exist is much larger than the asteroid, itself, hence the width of the track which can profitably be searched is much larger. Van Flandern et al. (1979) point out that a stable satellite orbit could exist out to ~100 times the asteroid's diameter. A more stringent range can be inferred from the fact that no asteroid has been observed to be a visual binary, implying that large satellites do not generally exist with separations greater than ~20 times the diameter of the primary. I have estimated (Harris 1979; see also Binzel and van Flandern 1979) that if asteroids have satellites at all, they should be $\geq$20 km in diameter (smaller ones would be destroyed by collisions with other asteroids), and would likely be found at separations of 5 to 10 diameters of the primary (due to effects of tidal friction). Thus, one could reasonably search for secondaries from as far off the track as 10 times its width (1000 km or more), and one should record the light level of the asteroid/star for at least ten times the maximum predicted event duration before and after the predicted time. If the maximum predicted duration is 20 seconds, one should monitor the light level for 200 seconds before to 200 seconds after the predicted time. To be safe, 5 minutes on either side would be a good choice. Since theoretical predictions are notoriously bad when based upon essentially no data, one might justifiably search even farther afield, but I think the above ranges serve to indicate the general requirements. In summation, one might profitably monitor the apulse of any event for which the predicted track passes within 1000 km of the site, which should provide a half dozen or so observing opportunities per year.

III. OCCULTATION PREDICTIONS

Unlike the predictions of occultations of stars by the Moon, one is faced with the perplexing difficulty that the uncertainty in the position of the occultation path is generally larger than the width of the path. An angular uncertainty of only 1/2 arc-second corresponds to a linear uncertainty at the distance of the asteroid belt of $\sim$1000 km. This same linear uncertainty is projected onto the Earth by the occultation geometry. The initial prediction uncertainty is a result of typical catalog errors in star positions and also similar scale errors in asteroid orbital elements. The only sure method of improvement is to photograph the asteroid and star on a single plate as they approach one another. One can expect an improvement to about ±0.2 arcsecond several days to a week before an occultation and perhaps ±0.1 arcsecond from plates taken the night before the occultation. Hence, in perusing a table of occultation predictions, one should pay attention to events that pass as much as 1000 km farther away then one is actually willing to travel, since the actual path may change by that much, or even more.

The limitations of the prediction process in general dictate the observing strategy. From predictions based on nominal star positions and orbital elements, events a year or more in advance can be identified which are unusually favorable for reasons of geography, expected accuracy of final predictions, scientific importance of the target asteroid, etc. Since the astrometric observations required to produce the essential final occultation predictions are beyond the capability of amateur (or even most professional) astronomers, one can only keep informed and wait for this work to be done. When the first update becomes available, a week or so before an event, it is possible to plan travel and site locations, or reject the event as impractical if the prediction moves the wrong way. This is the most important decision point because it is at this time that various individuals or teams declare their intentions and, ideally, agree upon one coordinate plan. If a "go" decision is reached, teams of observers will actually travel to nominal sites and await word by telephone of last minute updates of weather, and of track position based on plates taken the night before. Again, coordination is essential, since it is useless for everyone to run to the center line of the newly predicted path. It is essential to maintain an array of observers with enough breadth to assure uniform coverage over the range of uncertainty of the final prediction.

Occultation predictions based on nominal positions are published for the year in Sky and Telescope each January by David Dunham. Updated information can be had from him through the International Occultation Timing Association (IOTA) at the address and/or phone numbers given in those articles. Persons who have in the past coordinated photoelectric observing efforts are myself, Robert Millis of Lowell Observatory, William Hubbard of the

University of Arizona, and James Elliot of MIT. Any of the above may be contacted by amateurs equipped with photoelectric systems who desire to participate in occultation photometry.

IV. OBSERVATIONAL METHODS

The most important aspect of occultation observing is coordination, from beginning to end. The first step requiring coordinated agreement is the selection of an event to observe. Many more events occur than can possibly be observed adequately, and hence, only the best should be selected. This is confused by the uncertainty in advance predictions: what appeared to be an excellent event a year in advance may become less attractive as updated predictions become available, or vice versa. To aid in the assessment of event quality, Millis and Elliot (1979) define a quality factor, Q, as follows:

$$Q = 2 \sigma \Delta / d,$$

where σ is the angular uncertainty in the predicted position of the asteroid relative to the star, in radians, Δ is the earth-asteroid distance, and d is the diameter of the asteroid. Q is the ratio of the width of the zone of uncertainty to the width of the track. The probability that an observer, sitting on the predicted centerline of the track, will actually see an occultation is $\sim 1/(Q+1)$. The factor Q can be used two ways. By evaluating Q using the nominal uncertainty of advance predictions (typically ~ 1 arcsecond $\sim 1/200,000$ radians), one can estimate how far a predicted track might be expected to move upon refinement and thus compile a list of events which have a reasonable chance of becoming interesting, even though some of the nominal tracks may be unsatisfactory. The second use of Q is in planning the observing requirements. In this case, an uncertainty σ of ~ 0.2 arcseconds (10^{-6} radians) should be assumed in evaluating Q, as the uncertainty which will remain at occultation time. The final astrometry might be better, but should not be counted upon. In order to obtain n chords across an asteroid, the number of observation sites, N, needed is

$$N = n(1+Q)$$

Unfortunately, since smaller asteroids are generally more irregular in shape than larger ones, it becomes necessary to obtain more chords across a smaller asteroid in order to adequately measure the profile. Thus, for smaller asteroids, both n and Q tend to be larger, and the number of sites required skyrockets. Also, as Q increases it becomes more and more difficult to convince individual observers of the worth of their efforts. For a typical occultation, Q is ~ 2. While some researchers have accepted 3 as a minimum useful number of observed chords, I feel uncomfortable with less than 5. Thus, I would recommend ~ 15 observing sites as necessary to cover a typical occultation adequately. In the interest of conserving resources, it is important to mount efforts only for events where success is likely. Thus, if N computed from the second equation exceeds the number of sites that can realistically be mustered, then the event should be abandoned, except for monitoring for secondary events (Section II).

One must also consider geographical and meteorological aspects in selecting events. Predicted tracks have a habit of paralleling coastlines or international borders and then being updated across the forbidden line.

I provide the above information mainly for background. The actual selection and initial coordination of observing plans is best left to a very few professional astronomers, since they are ultimately the consumers of the data obtained; and also to avoid "calling wolf" too often. This last point cannot be overstressed. In the last few years, about 30 occultation events have been attempted, with many partial and inconsistent results and only a few clear successes. In order to maintain credibility (both with the observers and with government funding agencies) we must improve our record. For the near future, it is essential to commit all-out efforts to a few very favorable events and let the rest pass.

I have assumed so far that all observers are photoelectrically equipped. What, if any, is the role of visual observers? While I am a firm advocate of the photoelectric method, I confess I have attempted a few occultations visually myself. First, a visual observer can participate in the fun, mentioned at the beginning of the chapter. In addition, visual observers can contribute something to the scientific results. Experience has shown (Millis and Elliot 1979; Millis et al. 1981) that (a) in spite of careful assessments of "personal equations," visual timings are generally in error by 1/2 second or more; and (b) visual observers frequently report "events" where none probably occurred, thus the use of visual methods in searching for secondary events is of little recognized value. The best use of visual observations is in defining the edge of the occultation path. Strings of visual observers along each flank of the path can very effectively define the extreme width of the path, independent of timing accuracy. In planning observing locations, visual observers should position themselves along the edges of the expected track. If, as a visual observer, you choose to position yourself near the predicted centerline of an occultation, you will have minimized your chance to make a scientific contribution. While I cannot advocate against this strategy for a volunteer observer, you should be so advised.

As soon as an updated prediction is available (usually about a week before the event) a final "go" or "reject" decision can be reached. Past experience seems at first to verify Murphy's law: updated predictions are invariably less favorable than the nominal track. I suspect that this is actually a result of a myopic selection of only the "best" events for updating. The refined paths are bound to be worse. In the future, more prospective events should be considered, if possible, so that cases where the nominal track is unsatisfactory but the actual track is favorable will not be missed. Probably the major reason for the "poor" record of past efforts is the failure to reject once favorable prospects in the light of revised, unfavorable predictions, or conversely to mount hasty expeditions when last minute updates reveal a previously unexpected favorable opportunity. Both temptations should be avoided, as they have never paid off.

As soon as it is agreed to do an event, based upon an updated prediction, it is possible to assign sites to observers. The most successful technique which has been applied to date is to assign each observer to situate himself along a specific line parallel to the predicted centerline of the occultation. The parallel lines are spaced equally across a range of at least (Q+1) times the track width in order to assure that the actual path will be covered from side to side. Past experience has indicated that undue optimism has prevailed in evaluating Q, with the result that many events have been only partially covered, or missed entirely. I recommend a span of (2Q+1) times the track width, if possible. Perhaps a better strategy is to space the lines at 1/n times the track width, to assure the minimum of n required chords, and continue until the number of available observers is exhausted. In the unlikely event that a width of (2Q+1) is reached with more observers available, then additional observers can be interspersed between the initial set, with preference given to increasing the density of observers along the expected edges of the track.

As final updates of weather and track location become available, reassignments of observers may be in order. In order to do this, it is essential that an agreed upon coordinator remain in telephone contact with observers at all times before the event. Likewise, it is well to schedule check-in times by observers, since they will generally be otherwise out of contact. Last minute, rushed moves have proven to be unwise. It is best to formulate a strategy in time to allow a few hours excess time, and then stick with it. An individual observer is free to move to a more favorable location along the assigned track, however if weather forces a move off of the assigned track, it is imperative to check in with the coordinator as quickly as possible for reassignment and to allow the gap to be filled by an observer elsewhere.

It is important for the coordinator to keep track of the weather. Probably the best source is the network of FAA flight service stations at most commercial airports. Strictly speaking, their service is limited to pilots, however most of them will respond to a reasonable request for weather information for scientific endeavors. A second source which I have found useful is a local television news bureau. KABC television in Los Angeles has kindly allowed me access to their satellite weather pictures, which have aided considerably in siting observers on two occasions (for the Juno event shown in Figures 8-1 and 8-2, in particular). Other stations may respond similarly, especially if they perceive the event as newsworthy.

The task of coordinating an event can be subdivided (i.e., "You cover the north, we'll cover the south"), however in the past this has had the general result of redundant coverage over an inadequate width, since each coordinator wants his/her team on the centerline. Human nature being what it is, this problem will no doubt be with us forever. A better strategy is for each of two teams to take alternate lines across the track, preferably geographically separated along the track, so as to best guard against bad weather. If both teams are successful, the data will be complementary rather than redundant.

The actual technique of observing is fairly obvious from the discussion in Section II. One may be nagged by doubts that the correct star is being observed. The best cure for this is to begin soon enough to see the star and asteroid as a resolved double, if at all possible. This is usually not more than 20 minutes, depending on the motion of the asteroid, seeing, and magnitude difference. A tape recorder for making "notes" is helpful, especially if you are working alone. Call out any occurrence which could possibly affect the photometric record: guiding corrections, partial clouds, wind gusts, etc. The uncertainty in prediction time is Q times the maximum event duration, which will generally be only 10-20 seconds. Visual observers will do well to concentrate only on the primary event, perhaps only one minute total, since concentration for a longer time will likely degrade performance, and visual observations or secondaries are of limited scientific value. (It does no harm to continue looking after the predicted event, of course.) Photoelectric observations should span a time of at least 20 times the predicted maximum duration of the event, preferably twice that. However, extending the duration of coverage should not be allowed to interfere with the quality of coverage. Remember that uncertain results are of no value in the search for asteroid satellites and its not worth risking guiding errors at the critical time of the primary event.

The final step in occultation observing is reporting. If an event is recorded, the exact times of ingress and egress should be noted, with error bars, and the total range of time observed should be reported, with any breaks in the record (e.g., due to clouds, pointing problems, etc.) duly noted. It is important to report the range of time observed, even if no event is observed, in order to assess the completeness of the search for satellites. This includes observers outside of the track, of course. For observers near the track who recorded no primary event, that report is important in order to constrain the edges of the path. (Note the two "nearest misses" plotted in Figure 2.) In addition to the above data, the location of the observing site (to within 1 km or less), including the elevation above sea level, is necessary. If you do not have access to topographic maps, an odometer measurement to the nearest road intersection, marked on a road map, will do.

This data should be reported to the coordinator, to IOTA, or to the IAU Central Bureau for Astronomical Telegrams, Smithsonian Astrophysical Observatory, Cambridge, MA 02138. It is important to let others know of your observations in order to avoid multiple publication of partial results rather than a single publication of a complete result. If you wish to retain control over your data in order to assure proper credit in its eventual publication, this is perfectly acceptable scientific practice, but it is essential that you promptly report something, so that all the data can be eventually assembled. A tactful way to do this is to submit a preliminary report, not specifying exact location or event times, pending final reduction of the record. The question of propriety is generally unimportant for amateurs, but unfortunately is more so among professional astronomers where success in future funding often depends on demonstrated performance.

The final reduction of all of the data is not difficult. It consists of projecting the times and locations of events back onto the plane of the sky to form a profile of the asteroid. In addition, some estimate of the third dimension may be gleaned from the lightcurve, and the phase in the lightcurve at which the occultation occurred. Since occultation targets are generally among the largest and brightest of asteroids, the measurement of their lightcurves near the time of occultation is a useful project for a small observatory (see chapter by Binzel). The paper by Millis *et al*. (1981) is a good example of a final report of an asteroid occultation.

ACKNOWLEDGMENTS

Figures 8-1 and 8-2 are reproduced courtesy of R. Millis and the American Astronomical Society. The preparation of this chapter was supported at the Jet Propulsion Laboratory, Caltech, by NASA's Planetary Program under contract NAS7-100.

REFERENCES

Binzel, R. P. and van Flandern, T. C. (1979). Minor planets: The discovery of minor satellites. Science 203, 903-905.

Harris, A. W. (1979). The dynamical plausibility of asteroidal satellites. Bulletin American Astronomical Society 11, 560.

Millis, R. L., and Elliot, J. L. (1979). Direct determination of asteroid diameters from occultation observations. In Asteroids (T. Gehrels, ed.), University of Arizona Press, Tucson, pp. 98-118.

Millis, R. L. and 37 other authors (1981). The diameter of Juno from its occultation of AG+0°1022. Astronomical Journal 86, 306-313.

van Flandern, T. C., Tedesco, E. F., and Binzel, R. P. (1979). Satellites of asteroids. In Asteroids (T. Gehrels, ed.), University of Arizona Press, Tucson, pp. 443-465.

9. LUNAR OCCULTATIONS

Graham L. Blow

I. INTRODUCTION

A lunar occultation occurs when the Moon, travelling in its path about the Earth, passes in front of another celestial object. In the special case of a solar eclipse this object is, of course, the Sun, although it could be a planet, or most frequently a star. Occultations by the Moon are in fact among the oldest recorded astronomical phenomena; a lunar occultation of Mars in 357 B.C. for example, demonstrated to Aristotle that that planet lay further from us than did the Moon. Subsequent centuries saw other observers also use occultations to make deductions about the nature of celestial objects. Perhaps the most important of these was the realization that because stars, when occulted, disappeared more or less instantaneously, then their angular size must be very small.

In 1908, P. A. MacMahon suggested that if the time taken for the Moon's limb to occult a star could be measured, a direct value for the star's angular diameter might be obtained. MacMahon did not realize, however, that starlight passing close to the edge of the Moon would be diffracted--a fact quickly pointed out by Arthur Eddington. Eddington, himself, failed to realize that far from invalidating the idea, the shape of the diffraction pattern itself could provide the required quantity. The study of occultations for this purpose, then, languished for another 30 years. Interest was revived in 1939 by J. D. Williams, who calculated some sample diffraction patterns for light sources of very small angular extent. About the same time, A. E. Whitford used the 100-inch telescope at Mt Wilson (together with a moving film camera photographing the screen of a cathode ray tube) to actually record occultations of β Capricornii and γ Aquarii. Diffraction fringes were observed, but not in sufficient detail to permit the angular diameter of the stars to be determined.

The first true high-speed occultation observations in the modern sense came in the early 1950's, when D. S. Evans and A. W. J. Cousins in South Africa recorded a series of occultations of the red supergiant Antares. They found that its diameter of 0.040 arcseconds almost completely smoothed the expected diffraction fringes into what would be expected from geometrical optics alone. It was not until the late 1960's, however, that a regular photoelectric monitoring program on lunar occultations was begun by Evans, by then at the University of Texas, in collaboration with R. E. Nather. This led to an excellent series of papers fully describing the occultation process (Nather and Evans 1970), necessary instrumentation (Nather 1970), the effects lunar limb irregularities might have on traces (Evans 1970), methods of data analysis (Nather and McCants 1970) and observational results up until that time (Evans 1971). This program has continued uninterrupted to the present, and almost 6000 occultations have now been observed at the University of Texas' McDonald Observatory. Slightly more than one percent of these have allowed the angular diameter of the star involved to be measured. As a consequence of this work other observatories have been motivated to enter the field, and several groups elsewhere in the United States, Canada, Europe, and New Zealand are now involved observationally. Traditionally, because of the instrumentation requirements necessary to this type of observing, work has been largely restricted to the professional observatories. However, with the very low cost high-speed recording systems becoming available today (see following chapter), it may be that within a few years professional observers undertaking this work will be outnumbered by amateurs.

II. WHY OBSERVE OCCULTATIONS?

The observation of stellar occultation by the Moon, if recorded in the right way, can provide a great deal of important information about both bodies involved.

The Moon moves in its orbit at an average speed of 1.023 km/sec, or about 0.549 arcseconds per second, as would be seen from the center of the Earth. This rate is only actually achieved when the Moon is on the horizon, because at other times a contribution from the velocity of the rotating Earth tends to reduce the apparent rate to around 0".4/sec. This is, of course, the mean apparent rate of the Moon's center and thus leading point on the lunar disk, but consider for a moment a star being occulted at some different position on the disk (Figure 9-1). If the position angle of the Moon's motion is Φ , and the occultation takes place at a position angle of θ on the disk, then the apparent rate will be reduced by a factor Cos β , where β is known as the contact angle and is defined as

$$\beta = \theta - \Phi$$

For illustrative purposes though, if we take the average rate as being about 0".4/sec, then timing an occultation to an accuracy of one millisecond (0.001 second) fixes the position of a point on the lunar limb to 0".0004 which, at the Moon's mean distance of 384,400 km, corresponds to some 80 centimeters! Such accuracy is phenomenal, and together with lunar laser ranging data (which

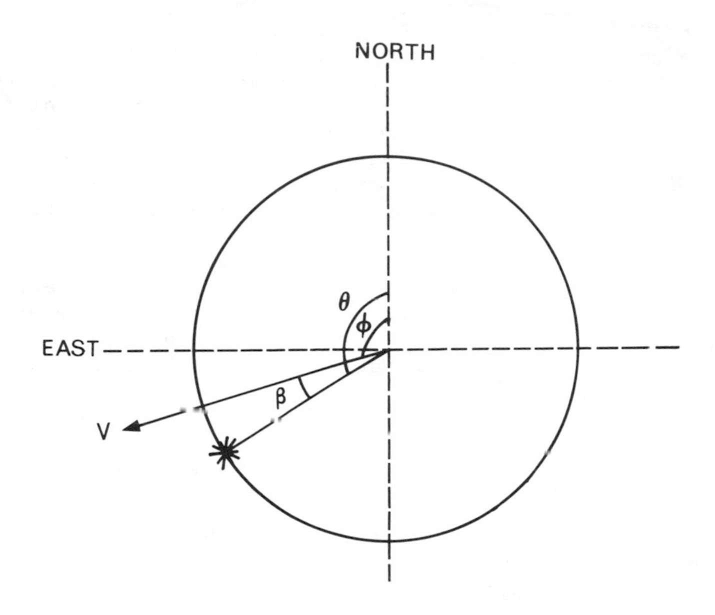

Figure 9-1. Determination of Occultation Time Scale.

determines the Moon's radial position to a similar or better accuracy), we could expect to define the position of our satellite in space very precisely indeed. However, things are not quite as simple as this, and because there are problems in adequately defining both the center of the Moon relative to any point on its limb, as well as the absolute position of faint stars, the overall accuracy is not as good as theory might suggest. Nevertheless, occultations still provide a more accurate means of determining the Moon's lateral motion than any other, and as such play an important part in deriving corrections to the lunar ephemeris, as well as providing an estimate of any secular acceleration of the Moon. From this point of view, the value of occultation timings increases as the years go by, since the uncertainty of the Moon's profile is going to be reduced by increasingly accurate satellite photographs of the lunar limb, while astrometry from space-borne telescopes (notably HIPPARCOS) will produce order of magnitude improvements in star positions.

As one might expect, the occultation method is ideally suited to the discovery and measurement of close double stars (with separations typically between 0".1 and 0".01). Such events are observed to occur in perhaps 10 percent of occultations, and it is then quite a simple matter to deduce the magnitudes of the two components, as well as their separation in the direction of advance of the lunar limb. It is worthwhile noting that a separation measurement of 0".01 is ten times better than can be obtained by visual observers under the best conditions. Furthermore, if the occultation can be recorded in two colors simultaneously, then otherwise unobtainable photometry of close doubles is possible.

Perhaps the most well-known application of the technique is in determining the angular diameter of stars--in fact, about 50 percent of all diameters determined so far have been obtained in this way. As mentioned in section one, the method involves measuring the visibility of the diffraction fringes at occultation and comparing them with those expected for a point source. The method works best for stars of late spectral type (since these are often observed to be giants), and is a complement to other techniques such as the intensity interferometer. Determining the angular diameter of a star is particularly important, as it has a bearing on several areas of fundamental astrophysics. In theory, if the angular diameter (Φ, in milliseconds of arc) and visual magnitude V (in the UBV system) are known, then the star's effective temperature T_e can be obtained through the relation

$$\log T_e + 0.1C = 4.2207 - 0.1V - 0.5 \log \Phi \qquad (9\text{-}1)$$

where C is the bolometric correction. Unfortunately, the quantity C is not very well defined, but T_e can still be obtained with an accuracy comparable to that of other methods of determination.

The right-hand side of equation (9-1) can be used to define a quantity F_V, which describes the visual surface-brightness of a star. T. G. Barnes and D. S. Evans (1976) have shown that there is an excellent correlation between F_V and the V-R color index. This has come to be known as the Barnes-Evans relation and is being further refined through continuing diameter determinations. The relation is one of particular importance, since it holds for intrinsic variable stars at each stage of their pulsation. Hence, by applying the technique to Cepheid variables, for example, and comparing angular and linear diameter changes, it is possible to determine a completely independent distance scale for the universe.

III. LIMITATIONS

Because an occultation is an event uniquely defined in time and space, we might expect that this places some fairly severe constraints on the use of the technique. For a start, the Moon's orbit is inclined at about 5°09' to the ecliptic, and the ecliptic is inclined to the celestial equator at an angle of 23°26'. Because the two nodes where the Moon's orbit intersects the ecliptic move westward (making one full circuit every 18.61 years), then when the ascending node of the lunar orbit is near the spring equinox, the Moon will reach declinations of +29° and -29°. Nine years later, however, it will only reach +18° and -18°. So taking into account the large size of the Earth, over a period of 18.61 years occultations of stars within a band about 10° wide may be seen from somewhere on the Earth's surface. This is about 10 percent of the total area of the sky, and is the major limitation on the use of the technique.

Added to this, however, are a number of other difficulties. The Moon's apparent diameter is about 1/2° and its daily motion about 6°, so that (knowing the average density of naked eye stars in the sky) we would expect to have an occultation of a naked eye star every one or two days. However, some events will occur in daylight, and hence be unobservable visually, and on average about half of the remaining events will be wiped out by adverse weather. In addition, occultations will not be easily visible for a few days either side of New Moon (when the Moon is close to the Sun) or Full Moon (when scattered moonlight interferes substantially). If this isn't enough, before Full Moon stars will disappear at the dark limb and reappear at the lighted edge and thus be effectively unobservable, while after Full Moon the situation is reversed, with disappearances at the lighted edge and reappearances at the dark limb. The latter, of course, are particularly difficult to record, since there is then the problem of trying to make sure the star will reappear in the small photometer diaphragm--this requires some form of offset guiding on a nearby star, or a telescope drive sufficiently good to keep the star accurately centered for the one hour or more that it might be behind the Moon.

Given these difficulties one might wonder how occultation observing ever got started at all! However, the frequency of events of moderate brightness ($m_v \leq 9$) is such that there is still great scope for work in the field.

IV. OCCULTATIONS AND DIFFRACTION

In order to appreciate the remainder of this chapter fully, it is important that the reader have some understanding of the diffraction phenomenon. So it is worthwhile digressing for a short while to discuss diffraction as it pertains to occultations.

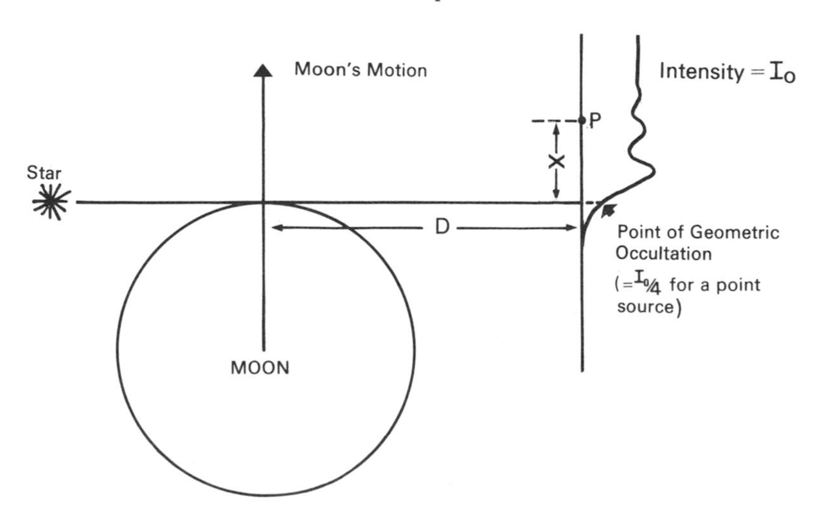

Figure 9-2. Diffraction at the lunar limb.

Point Source Stars

Figure 9-2 illustrates the situation. As the Moon passes in front of a star, a series of alternating bright and dark fringes will sweep across the observer's telescope. The brightest of these fringes will be closest to the point of geometric occultation (the 25 percent intensity level for a point source), while the other fringes decrease in both spacing and amplitude. At the Moon's distance the initial fringe spacing is in the order of 10 meters, and their frequency of oscillation about 40 Hz.

The light intensity at any point P, a distance x (Figure 9-2) from the geometric occultation point, can be described by

$$I(\omega)_P = \frac{I_o}{2} \quad (0.5 + C(\omega))^2 + (0.5 + S(\omega))^2$$

where $C(\omega)$ and $S(\omega)$ are the Fresnel diffraction integrals, I_o is the intensity a long way before the occultation point, and the dimensionless Fresnel number is given by

$$\omega = x \left(\frac{2}{\lambda D}\right)^{1/2}$$

So the intensity at point P turns out to be a function of both the wavelength of observation, λ, and the lunar distance D. However, we do not observe an occultation in one wavelength only, but rather over a spectral region whose bandwidth is determined by the spectral distribution of the star's light and the sensitivity of our detecting system to different wavelengths. In practice we can approximately allow for this by dividing our region of interest (usually the visual region, λ = 300 nm to 700 nm) into finite segments (or "optical samples") of width say 10 nm. Then the power P, radiated by the star in each segment i, can be calculated from

$$P_i \quad (\lambda_i) = \int_{\lambda_i - 5\mathrm{nm}}^{\lambda_i + 5\mathrm{nm}} B(\lambda) \quad d$$

where λ_i is the central wavelength of the segment and $B(\lambda)$ is the radiated flux from the star as described approximately by the well-known black body formula. This distribution is going to be reddened by both interstellar and atmospheric absorption, although in practice the former can generally be ignored since the majority of stars which will be bright enough to produce a perceptible diffraction pattern will lie within a few hundred parsecs. The second factor is easily taken into account by dividing the atmospheric transmission curve (Allen, 1973) into the same i segments, and defining A_i to be the transmission of the atmosphere in the ith segment.

Similarly, the components of the detecting system (photomultiplier (T), filter (F), and mirror (M)), each have a specific spectral response, so that the total spectral response in any segment i will be

$$C_i = P_i A_i T_i F_i M_i$$

The expected diffraction pattern I(x) will now be the sum of the individual weighted monochromatic patterns. Thus

$$I(x) = \sum_i C_i I(\omega_i)$$

(where C has now been renormalized). Experimenting with this shows that the shape of the final curve is very dependent on bandwidth. This is illustrated in Figure 9-3, where patterns computed for bandwidths of 0, 200, and 400 nm, all centered on λ = 500 nm, show a successive decrease in detail among the higher order fringes. In particular, the maximum intensity of 137 percent of the pre-occultation value only occurs for the monochromatic point-source case.

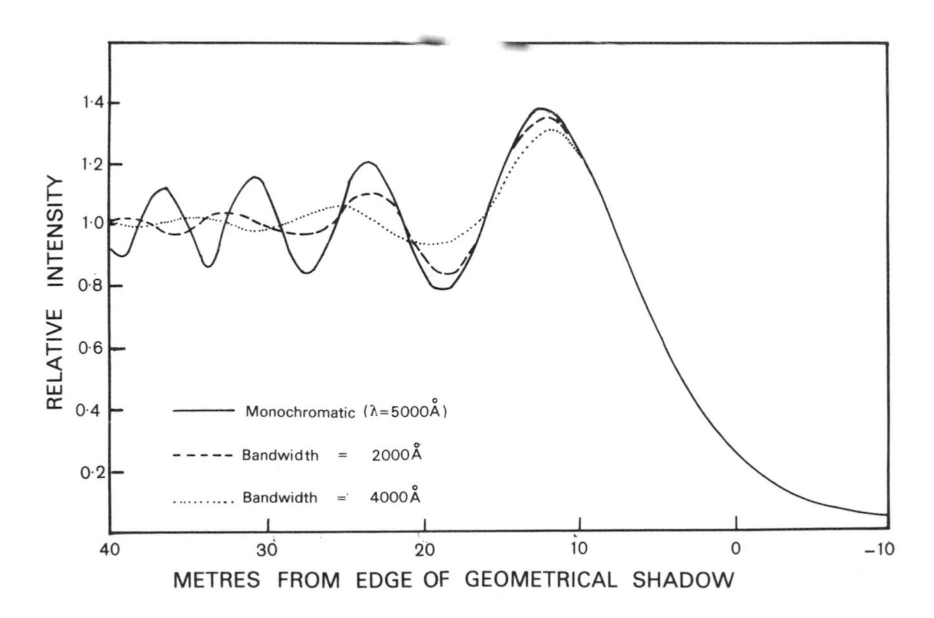

Figure 9-3. Effect of increasing optical bandwidth on the diffraction pattern.

Putting this another way, the narrower the optical bandwidth is made, the better the fringe resolution that can be achieved. Since the only way we can govern the bandwidth at the telescope is by changing filters, then clearly, using filters with a very narrow bandpass should achieve excellent resolution. However, the problem of actually receiving enough light through the filter to produce an acceptable counting rate then becomes significant, and so in practice a compromise is always necessary if reasonable resolution, at an acceptable data level, is to be achieved. It is only on the very brightest stars that interference and other narrow-band filters can be used to best effect.

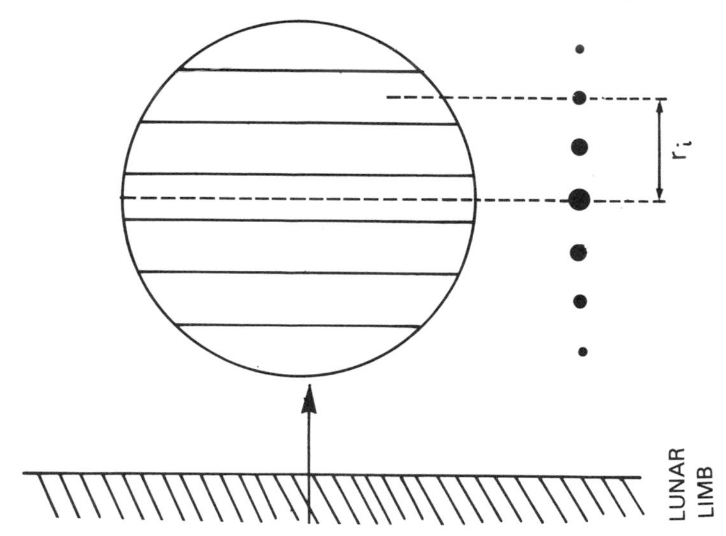

Figure 9-4. Modeling a stellar angular diameter.

Finite Diameter Sources

The model constructed above can be extended to take account of stars with a perceptible angular diameter d, as follows. If the star's disk is divided into a series of n strips of equal width, aligned in the direction in which the lunar limb is advancing, then each strip can be approximated by a point-source whose intensity is weighted according to the normalized area of that strip. The situation is illustrated in Figure 9-4. The angular separation between point sources will be $\frac{d}{n}$, which at the Moon's distance D corresponds to a distance $D\frac{d}{n}$ on the Earth. So, the total diffraction pattern one would expect will be given by a linear superposition of the n point-source patterns, each weighted and separated by a distance $D\frac{d}{n}$. Limb-darkening can be taken into account by altering the weight assigned to each point source. However, going from a non to fully limb-darkened disk only changes the derived diameter by about 12 percent.

Figure 9-5 shows the effect that increasing stellar diameter has on a (monochromatic) diffraction pattern. The most obvious change is that the fringe amplitude progressively decreases, and the fringes become more spread out, until for stars greater than about 0".012 the curve rapidly approaches what we would expect from purely geometrical optics; i.e., a smooth decline in intensity with diffraction playing no part. A perhaps less noticeable change is that the intensity at the point of geometric occultation changes as a slow function of stellar angular diameter.

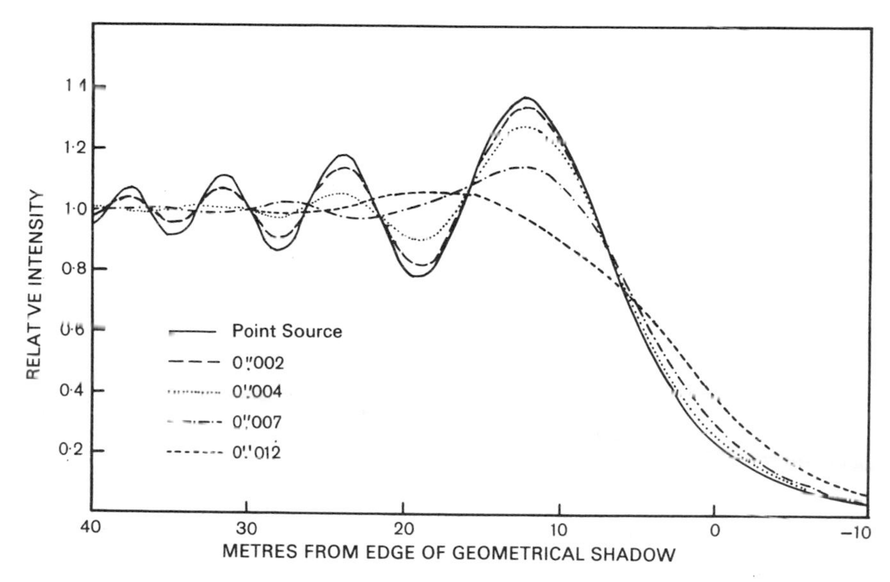

Figure 9-5. Effect of increasing stellar angular diameter on the diffraction pattern.

Obviously, there will be some limit to how well a star's angular diameter can be resolved, and this will be dependent on the visibility of the diffraction fringes. As mentioned earlier, the optical bandwidth we allow for observing the occultation is going to affect fringe resolution. This turns out to be the most serious restriction, and experiments by Nather and McCants (1970) have shown that a typical bandwidth of 100 nm will limit resolution to about 0".0025. The telescope aperture will also have a similar effect, since for large telescopes one edge of the mirror will see the occultation one or two milliseconds before the opposite edge does. This will tend to smear out the pattern somewhat, although it can be accommodated by considering the mirror as a series of slices, where each slice sees the same pattern at a slightly different time--analogous to the case of the stellar strips. However, taking this into account is not really necessary for small telescopes; for example, a 60 cm telescope will only limit resolution to 0".0003. Other factors to be considered are the integration time (two milliseconds imposes a limit of about 0".0008) and the width of the optical sample used in calculating the theoretical pattern. (The width of 10 nm used in the previous section will only limit resolution to about 0".0005).

All the above factors are ones over which the observer has some degree of control. However, atmospheric turbulence, which comprises both astronomical seeing (tending to both smear the star image out, and randomly displace it), and intensity scintillation, are going to have some additional effect. In very crude terms, the former can be regarded as increasing the telescope aperture by a few centimeters, while the latter will tend to increase or decrease the amplitude of the diffraction pattern.

V. EQUIPMENT

The equipment necessary for successful photoelectric observation of lunar occultations is discussed in detail in the following chapter, so the following should be regarded only as a summary.

Almost any photometer, whether in current to frequency or direct photon counting mode, is suitable. However, the main requirement is an acquisition system with the ability to gather and store data at very high time resolution--typically integration periods of one to eight milliseconds. In most modern detecting systems data acquisition at this rate is carried out using a block of computer memory as a storage buffer. For occultations this buffer is "rotating," so that the most recent record of the star's intensity is successively replacing the earliest remaining record. Thus, for a 4K block of storage, gathering data in one millisecond bins, a continually updated history of the star's intensity over the preceding four-second period is maintained. Some means of continuously examining the contents of this buffer is highly desirable, and this is best accomplished with an oscilloscope display, so that when a drop on the screen corresponding to the event is observed, the rotation of the buffer can be stopped and the contents "frozen." If the computer program can record (to millisecond accuracy) the time at which the rotation was stopped, then it will be a simple matter to count back from the end of the buffer and determine the absolute time of the event. In past years minicomputers which have been used in this way were somewhat cumbersome PDP-8s, Novas, etc., but microprocessor-based devices produced in recent years will carry out the same functions at a fraction of the cost.

As the above implies, part of the acquisition system needs to be an accurate clock which is either in synchronism with absolute time, or at some known offset from it. This can usually be arranged by comparing the clock with a signal from one of the shortwave time service stations (WWV, for example). One should beware, however, that the radio signal takes a finite amount of time to propagate from its source to any receiver, so that in general, clocks should either be set to take account of this delay, or the time delay later applied as a correction to the deduced time of the occultation. (Hewlett-Packard Applications Note No. 52 (1965) gives information on determining propagation delays.)

It is rare for any clock to measure time as accurately as one would like, and most will drift at some rate which can be measured by repeated comparisons with a standard. Knowing the rate of drift and the times at which comparisons were made, the clock error can be interpolated for the time of the occultation, and also applied as a correction to the deduced time of the event.

A further desirable piece of equipment is some sort of storage device, so that data can be stored for later analysis at leisure. This could be via floppy disks, or perhaps cassette or paper tape units, although these tend to be alternately costly, unreliable, or hard to obtain. Perhaps the best storage method for those on a limited budget is to simply obtain a listing of the data on an ordinary teletype or printer. Although this inevitably means a lot of numbers on paper, they are nevertheless quickly available at any time for future reference.

Once an occultation system has been assembled and functions reliably, there are several modifications that can be made to improve the data gathering process. The first of these is by cooling the photomultiplier tube so that fainter events can be detected. In some cases the signal-to-noise ratio increases dramatically, but caution should be exercised as not every type of tube is amenable to this sort of treatment. See Chapter 4 of Hall and Genet (1982) for a fuller discussion of this.

A further modification is an offset guider which, when mounted in front of the photometer, will enable reappearances to be observed. A dual-channel photometer, permitting simultaneous monitoring of an occultation in two colors, has advantages in both double star work and in picking out cases of distortion in the trace due to lunar limb irregularities. However, both these approaches tend to involve some considerable extra expense and are not really necessary in the short term.

VI. OBSERVING TECHNIQUES

In order to successfully observe occultations there is one essential item which has not yet been mentioned--predictions! These can be computed for any location by the

Nautical Almanac Office
U. S. Naval Observatory
34th and Massachusetts Ave., N. W.
Washington D.C., 20390
U.S.A.

and are available gratis on an annual basis. Requests for predictions, however, should include information on the size of telescope to be used and its geographic coordinates, and photoelectric (as opposed to visual) option predictions should be asked for. As well as listing the predicted time, and magnitude, position and spectral type for occulted stars, the predictions also contain information essential in the reduction process and not easy to compute otherwise. This includes the distance of the Moon in kilometers (D) at the time of the event, and the radial rate of approach of the Moon's limb to the star (i.e., diffraction fringe velocity) in arcseconds per second; conversion to meters per millisecond is accomplished by multiplying by D/206265. Also given is the event's contact angle on the Moon, which is required before the slope of the Moon's limb at the point of occultation can be determined (see Section VII).

One column in the predictions that should be noted carefully is the "D", for double star, column. The code letter appearing here indicates that the star is double or triple, and approximately how close the components are. Particular attention should be paid to the A, C, and O codes, which generally indicate a visual double. M implies the star position given is the mean of the two components. In the case of these codes, looking up a double star catalog to obtain the approximate position angle and separation is desirable, since with the position angle (on the Moon) of the event given in the predictions, the separation of the stars in the direction in which the lunar limb is advancing can be computed. Using this information in conjunction with the listed radial rate of approach will indicate whether the time between occultations is greater than the circulation time of the storage buffer, in which case the integration time (and hence the rotation period of the buffer) should be increased. If a double star catalog is not available, the same information can be approximately obtained by using the known size of a photometer diaphragm, and estimating the position angle of the two stars.

For bright ($m_V \leq 5$) late type (K, M,...) stars it is useful to know whether an event has the potential to yield a resolvable angular diameter. For this, the Barnes-Evans relation can be used. From Barnes and Evans (1976)

$$F_V = 3.977 - 0.429(V-R)_o \tag{9-2}$$

where

$$0.0 \leq (V-R)_o \leq 1.26$$

Obtaining $(V-R)_o$, the V-R color index as seen from above the atmosphere, can be a problem, although since only an approximate value for ϕ is required, it is probably reasonable to use the measured V-R value given in some catalogs (e.g., Iriarte et al. 1965). Then, since F_V equals the right-hand side of equation 9-1, the visual magnitude V can be used to solve for ϕ .

For occultations of bright stars it is advisable to use some sort of filter. This has the dual function of cutting down the count rate (so that the phototube will not become overloaded), and enhancing the diffraction fringes, which in the case of red stars increases the chances of determining an angular diameter. If a Johnson UBV filter set is available, the use of, for example, a

B filter for O, B, and A stars, and a V filter for F, G, K stars, is advisable. For M and N stars a redder filter still might be preferred (although care must be exercised as the photocathode used will ultimately determine the red response cut-off).

In occultation photometry it is important to remember that one can be dealing with a sky background that is considerably greater than in other sorts of photometry. In fact, with an eight arcsecond photometer diaphragm placed at the midpoint of the dark limb of the first or last quarter Moon, scattered moonlight will be approximately equal to the light from a 5th magnitude star--so that an 8th or 9th magnitude star will only contribute a few percent to the signal. In dealing with high count rates then, in order to avoid damaging the tube it is best to start on a lower voltage and increase it gradually to a suitable operating level. This is particularly important in GaAs type tubes (see Hall and Genet 1982, Chapter 4).

The count rate and signal-to-noise ratio can also be controlled by a judicious choice of diaphragm size. Sizes from 60" down to 8" are useful, although obviously the larger sizes will tend to limit observation to the brighter stars whose signal will still be discernible above the increased background. In general, an observer should aim to use the smallest possible diaphragm which will not interfere with the observation in other ways. Final choice will depend on the lunar phase and consequent background brightness, and on the state of the atmosphere (since care must be taken to ensure that all of the star's seeing disk remains within the aperture boundary). Furthermore, for small diaphragm sizes the requirements of telescope tracking are increased, so that if a variable frequency drive is available it is often desirable to experiment first and get the tracking rate just right. In particular, the tracking rate for an object near the horizon is usually substantially different from the rate for the same object on the meridian, while at very low altitudes the drift in declination due to refraction is also noticeable. By the time this latter factor becomes significant though, the star is generally too low to yield a reasonable quality trace anyway. Telescope tracking should be checked right up to 60 seconds or less before the event.

Once data is being gathered, running the star in and out of the diaphragm will give some indication of the signal-to-noise ratio, and drop in level to be expected at occultation. If abnormally high or noisy data is obtained, first check that the photometer aperture illumination has been turned off--it is easy to forget!

At very low lunar phase ($\sim$ 20%) earthshine on the dark limb is often significant, and with the lunar limb intruding into the aperture just before the occultation this is sometimes noticeable as a substantial increase in noise. One should also be aware that at high phase the background can increase sharply as the event approaches, due to the proximity of the terminatory. On these occasions it is important that the photometer dark slide be put in as soon as the event has been captured. Of course, both the low phase and high phase problems can be assisted by prior choice of a fairly small diaphragm.

It is important to be ready to capture an event perhaps a minute before the predicted time, as occultations can occasionally occur up to 30 seconds before they are expected--this is in spite of the entry in the AC column of the predictions, which is intended to give a rough indication of possible errors in the predicted time. The accuracy quoted in this column is typically ± 3 seconds, unless the event occurs near one of the lunar cusps. After the event has occurred, wait a second or two before stopping the buffer, just in case the star is double. Be aware of this possibility before the main event as well, and look for small but definite, abrupt decreases in brightness, that do not sink to the level expected.

In the vast majority of cases the rotation of the buffer will be able to be stopped when the occultation is seen (either on the oscilloscope screen or through a guide scope), and the data stored with no problem. However, above all, don't panic if everything goes wrong ten seconds before the event. University of Texas observers claim to have found 57 ways to foul up an occultation, and this author feels that he has independently discovered a good number of these. In spite of the most fastidious preparations, disasters will happen to everybody sooner or later--so there is no point in panicking unnecessarily over them.

VII. DATA REDUCTION AND ANALYSIS

Occultation Times

Until now the discussion has mostly concerned occultations of stars which will yield a discernible diffraction pattern. However, for most occultations the background is so bright and the star so faint that this pattern is lost in the noise, and one ends up with a trace in which a plain drop in intensity indicates the event. Such an occultation was that of SAO 146861, observed at the McDonald Observatory on 1981 November 8, and shown in Figure 9-6. Even in this case, it is still possible to fit a point-source diffraction pattern to the trace and so determine the 25 percent point (which, as mentioned in Section IV, corresponds to the point of geometric occultation). Carrying this out for SAO 146861, and knowing that data collection was stopped at $02^{h}20^{m}48^{s}.496$ UT (with each data bin 8 milliseconds long), the time of occultation is easily determined as 1981 November $08^{d}02^{h}20^{m}44^{s}.931 \pm 0^{s}.003$, where the quoted error takes account of changing time signal propagation delays, and such like.

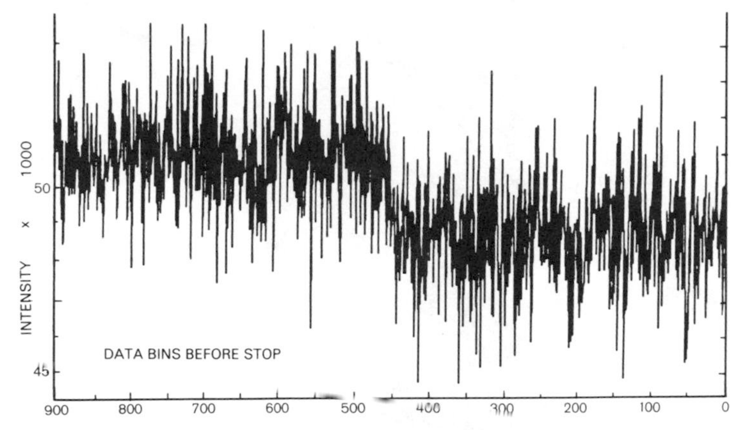

Figure 9-6. Occultation of SAO 146861, 1981 November 0.

However, fitting a diffraction pattern to every noisy trace can be a time-consuming process, and it is often easier to instead take the mean level between the pre- and post-occultation data as the occultation point. The error this introduces into the deduced time generally only amounts to a very few milliseconds, and is often less than errors due to other sources. As an aid to determining the occultation location it is sometimes useful to construct an _integral plot_ of the data, as first suggested by Dunham _et al_. (1973). This is a graph of the function

$$\mathrm{Int} = \sum_{1}^{n} (I_{x-n} - \overline{I})$$

versus n, where n is the channel (data bin) number, x is the total number of channels and $\overline{I}$ is the mean intensity over all channels. Figure 9-7 is such a plot for our example occultation, and shows that the effects of noise have been largely removed. The occultation point is immediately obvious as an abrupt change in slope.

Determining occultation times in this way is of little use unless they can be made available to others. At the present time, the global body which collects all occultation timings (whether photoelectric or visual) is the

International Lunar Occultation Center
Astronomical Division
Hydrographic Department
Tsukiji-5, Chuo-ku
Tokyo, 104, Japan

from which forms and other material to assist in reporting timings are available.

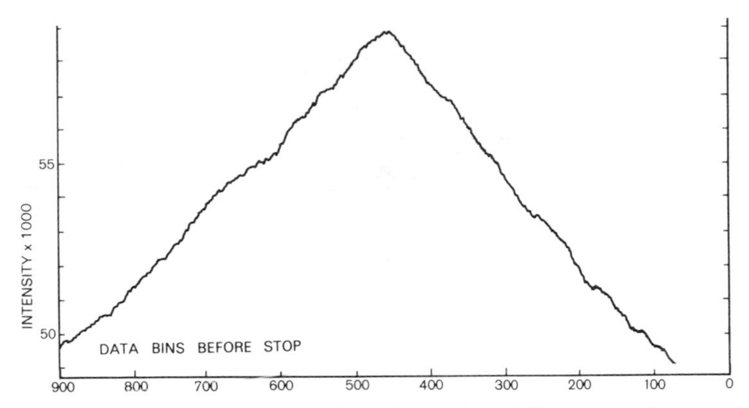

Figure 9-7. Integral plot for SAO 146861 occultation.

Double Stars

Occasionally, the occultation of a double star will be observed, in which case the starlight will usually drop to some intermediate level for a short period, before falling to its final value. If the star is bright and the conditions favorable, diffraction patterns will be produced by both stars, and the final observed trace will be the sum of these two patterns. More often, however, a trace such as Figure 9-8 is obtained. This is from the occultation of ZC 3015, observed (in blue light) from Mt John University Observatory, New Zealand, on 1979 October 1. The diffraction patterns are masked by noise, but the integral plot (Figure 9-9) clearly indicates where the occultations took place. Because the two steps are well-defined, it is possible to work out the magnitude of each occulted component of ZC 3015. From the basic equations relating luminosity and magnitude, the expression

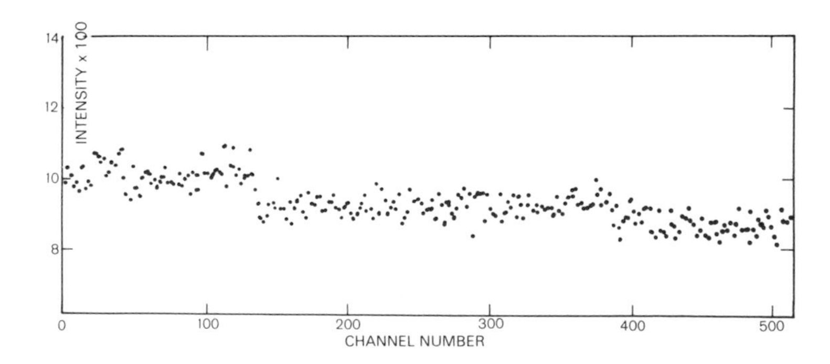

Figure 9-8. Occultation of the Double Star ZC 3015, 1979 October 1.

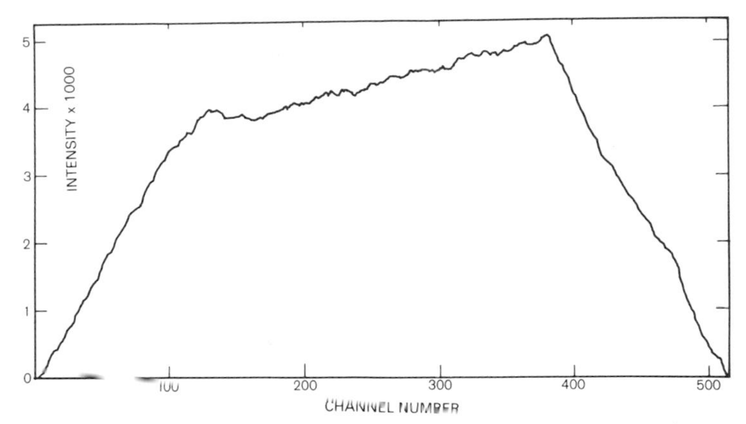

Figure 9-9. Integral plot for ZC 3015 occultation.

$$m_A = 2.5 \{ \log (\ell_A + \ell_B) + 0.4 m_T - \log \ell_A \}$$

can be derived for the magnitude of component A of the system (with a similar expression for m_B), where ℓ_A is the intensity drop from the initial to intermediate level, ℓ_B is the drop in intensity from the intermediate to final level, and m_T is the total magnitude of the system. For ZC 3015, by simply taking an average over the pre-, post-, and intra-occultation data we can determine that $\ell_A = 77.4 \pm 4.84$ and $\ell_B = 58.95 \pm 4.68$. Since the total magnitude is 5.30 (Hoffleit, 1964), the component magnitudes must be

$$m_A = 5.92 \pm 0.18$$

$$m_B = 6.21 \pm 0.22$$

The vector separation of the pair can also be calculated. This is the separation in the direction of advance of the lunar limb (and not the true separation which, together with the true position angle, can only be determined if the event is observed simultaneously from two locations with the star at different position angles on the lunar limb). Figure 9-10 illustrates vector separations as seen from two sites.

For our example star, the predictions give the radial rate of lunar approach as 0.4992 arcsec/sec so that, knowing that there were 126 bins of 4 milliseconds each between the events, then the vector separation is 0".252. The predictions also give the position angle of the event on the lunar limb (71°.1) and the contact angle β (-8°.7). Since we have already defined β as the angle between the position angle of the star and the direction of motion of the Moon, then the vector separation of 0".252 must be measured in a P.A. of 79°.8.

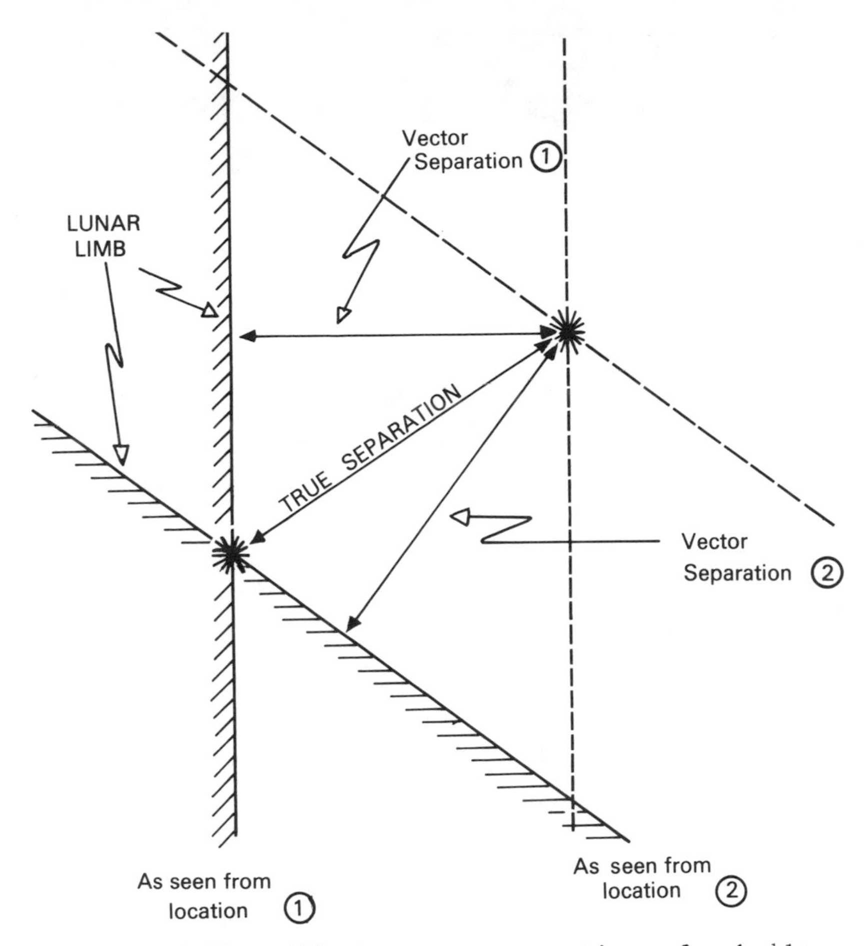

Figure 9-10. Differing vector separations of a double star, for the same occultation seen from two sites.

In just the same way as for SAO 146861, the times of occultation of the two components can be found:

$$T_A = 12^h34^m39\overset{s}{.}683 \pm 0\overset{s}{.}006 \text{ UT}$$

$$T_B = 12^h34^m40\overset{s}{.}187 \pm 0\overset{s}{.}008 \text{ UT}$$

ZC 3015 is listed in several double star catalogs, with A and B component magnitudes of 5.6 and 6.9, respectively. (The star is actually triple, but the third component is magnitude 12.1 and too far away to affect the results). An ephemeris published by Muller and Meyer (1969) indicates that at the date of occultation, the A - B separation would be $0\overset{''}{.}18$ in a true P.A. of $116\overset{\circ}{.}4$. Since the vector separation can never be greater than the true separation, this prediction (along with the magnitudes quoted) is at variance with the results.

The problem is resolved when one appreciates the enormous difficulties faced by visual double star observers in trying to measure stars less than one arcsecond apart. Confirmation of this comes from another occultation of ZC 3015, observed by Africano et al. (1975) at McDonald Observatory on 1974 August 30. On that date a vector separation of $0''.2566 \pm 0''.0002$ in a P.A. of $64^{\circ}.7$ was found, with a magnitude difference of $\Delta m = 0.42 \pm 0.05$ (in blue light). For the same time, Muller and Meyer had predicted, based on a visually derived orbit alone, a true separation of only $0''.22$ in a P.A. of 113°. This is a good illustration that one should not necessarily accept all published results as fact!

Point-Source Stars

The majority of stars bright enough to give a perceptible occultation diffraction pattern will also, unfortunately, have angular diameters too small to be measured. As an example of this sort of event we can use an occultation of the magnitude 5.17, A0 star ZC 2686, which was observed with the 61 cm Boller and Chivens reflector at Mt John University Observatory, New Zealand on 1979 September 29. The photometer contained an EMI 9558Q phototube and the event was monitored through a blue filter consisting of 0.5 cm of Schott BG14 glass. Figure 9-11 shows the total expected spectral response, taking into account the response of the phototube and filter, the reflectivity of the telescope mirror, atmospheric transmission and the spectral distribution of the star's light.

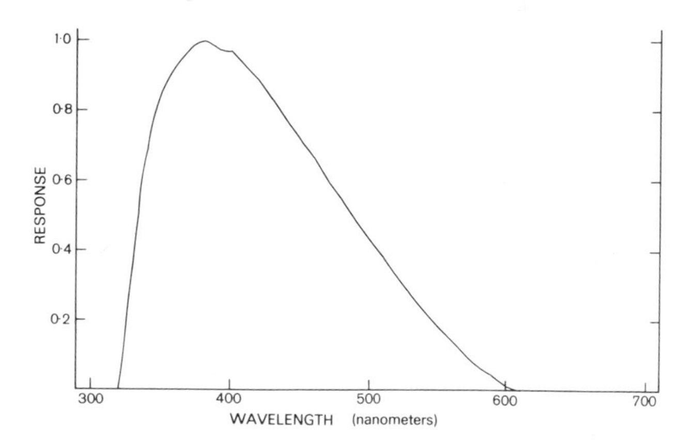

Figure 9-11. System spectral response for the occultation of ZC 2686.

The expected angular diameter for a star of this magnitude and spectral type can be roughly estimated through equations 9-1 and 9-2. Using $(V-R)_0 = (V-R) = +0.02$ (given by Johnson, 1966, for an A0 star), the angular diameter comes out as 0$''$.0003, which is too small to be resolved by the detecting system used (Section IV).

The points in Figure 9-12 show the data for ZC 2686, and are two milliseconds apart; the abscissa is plotted as meters from the edge of the geometrical shadow. The solid curve is the point-source diffraction pattern which can be best fit to these data. How was this curve obtained?

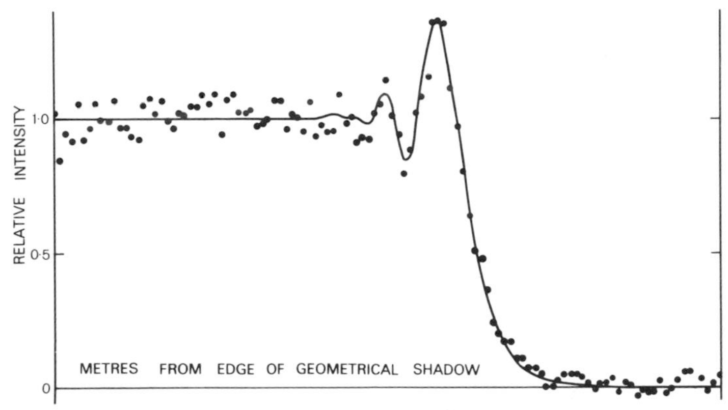

Figure 9-12.
Occultation of ZC 2686. Data points are 2.0 milliseconds apart.

Section IV covered the basic theory behind the calculation of a diffraction pattern. By breaking the calculated spectral response (Figure 9-11) into 10 nm segments and following Section IV's outline, we can calculate the expected diffraction pattern for this occultation and attempt to fit it to the data. The fitting process is an iterative procedure involving the four operations below, and illustrated in Figure 9-13.

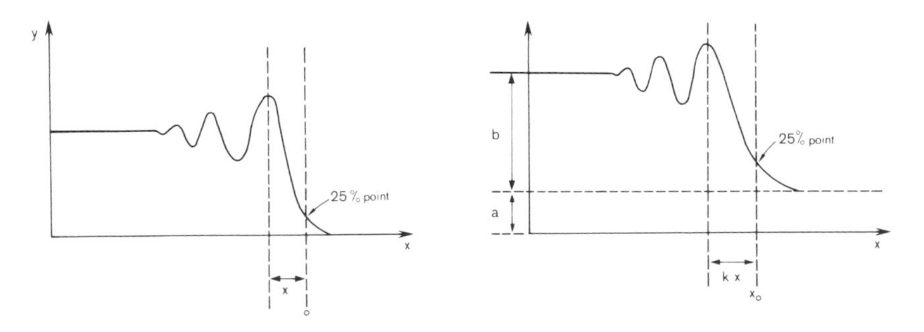

Figure 9-13. Operations involved in fitting a point-source diffraction pattern.

a) Displacement of the y scale = Sky background
b) Amplification of the y scale = Stellar intensity
c) Displacement of the x scale = Time of geometric occultation
d) Amplification of the x scale = Fringe velocity

Therefore, if the theoretically calculated diffraction pattern (which will be a plot of normalized intensity versus distance) is $I(x)$, then the fitted pattern $I(x)_F$ is given by

$$I(x)_F = b \{ I [k(x-x_o)] \} + a$$

where all the variables are as defined in the diagram. The four parameters can then simply be altered, and a new pattern calculated until a suitable statistical test (e.g., a chi-square test) indicates that the best fit has been obtained. Usually three or four such iterations are sufficient, and a published fitting algorithm is usually adequate (e.g., Bevington 1969).

Carrying out these operations for ZC 2686 gives a best-fitting curve with reduced chi-square 1.019, and the following parameters

Background = 145.75 ± 2.49
Stellar Intensity= 515.23 ± 4.06
Occultation Time = $07^h14^m49\overset{s}{.}691 \pm 0\overset{s}{.}006$
Fringe velocity = (0.5516 ± 0.0098) meters/msec

The USNO predictions for this occultation give a lunar distance of 371,877 km, with the event at a position angle of $129\overset{\circ}{.}0$ on the Moon. The predicted fringe velocity of $0\overset{''}{.}3000$/sec therefore translates to 0.541 m/ms, slightly different to the derived rate above; i.e., the actual diffraction pattern is compressed in time with respect to the predicted pattern. This difference can be attributed to the presence of an "average" slope with respect to the mean lunar horizon, over the perhaps 50 meters of lunar limb involved in forming the diffraction pattern. This is illustrated in Figure 9-14 where the slope is at angle α and the other angles are as defined in Section II. As mentioned in that section, the predicted rate of advance of the lunar limb in the direction perpendicular to itself, V_p, will be

$$V_p = V \cos \beta$$

while the actual rate V_a is going to be

$$V_a = V \cos (\beta - \alpha)$$

Hence, α can be calculated from

$$\alpha = \beta - \cos^{-1} \left\{ \frac{V_a}{V_p} \cos \beta \right\} \qquad (9\text{-}3)$$

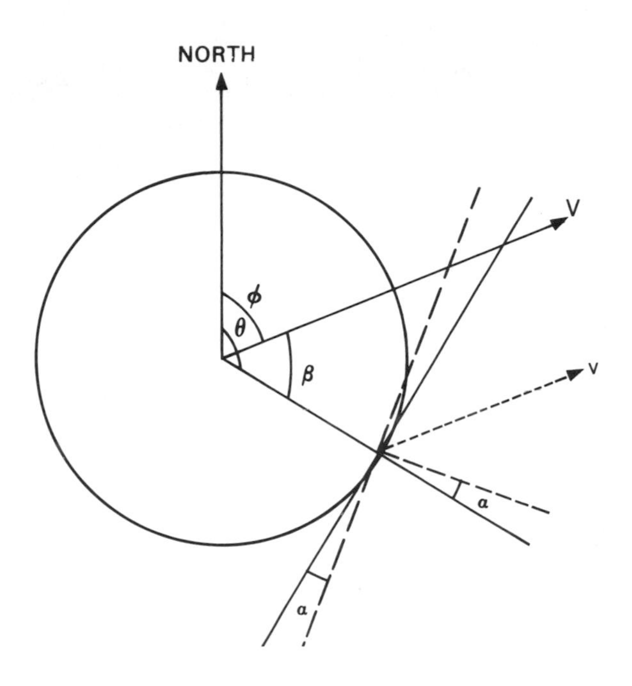

Figure 9-14. Determination of Lunar Slope α.

However, since Cos $(\theta - \phi - \alpha)$ = Cos $(\phi - \theta + \alpha)$, this actually gives rise to two solutions for α; one generally corresponds to a plausible slope, and the other to an extremely steep slope. The latter can usually be ruled out, as the Apollo missions have shown us that slopes on the Moon tend to be really quite gentle. Further, if we consider that during an occultation the light from a star will be skimming the lunar surface, the background hills will tend to fill in foreground valleys (and vice versa), making the lunar horizon appear quite smooth. This effect is easily verified with terrestrial horizons.

For ZC 2686, applying equation 9-3 gives a lunar slope of

$$\alpha = +1^{\circ}\!.32 \pm 0^{\circ}\!.69$$

where the plus sign is used by convention to indicate a compression of the pattern (A negative slope expands it). Occultation observations over many years have shown that most slopes on the Moon are $\leq 15°$.

Very occasionally, limb irregularities can play a much more important role. It is possible for very small scale bumps, of perhaps a few meters extent, to distort a diffraction trace by delaying part of the pattern relative to the rest. Any trace, for example, which exhibits a first peak more than 137 percent of the pre-occultation intensity (allowing, of course, for noise) must be distorted in this way, although on other occasions it is difficult to decide whether a strange looking trace is due to a limb irregularity or two stars so close together that their diffraction patterns are adding in an odd way. The only effective way to resolve this from one observing site alone is to monitor the event simultaneously in two colors, since the size of the first Fresnel zone is a function of wavelength, and thus the effects of limb irregularities will appear different in the different wavelengths. Fortunately, distorted traces from limb obstructions are rare and seem to occur less than one percent of the time.

Angular Diameters

On November 14, 1981 the reappearance of SAO 77516 ≃ Y Tauri from behind the Moon was observed (by offset guiding) with the McDonald Observatory's 76 cm telescope. Y Tauri is both an extremely red carbon star (spectrum C5III), and a semiregular variable with visual range 6.8 to 9.2 in a period of 240.9 days--at the time of occultation the star was near maximum. The event was monitored in both red and blue light, and the red data (obtained through a narrow-band RG610 Schott filter) is plotted in Figure 9-15. Data points are four milliseconds apart, and because the event was a reappearance, time runs from right to left. The first diffraction peak is substantially less than 137 percent of the pre-occultation value, implying that Y Tauri may have a measurable diameter.

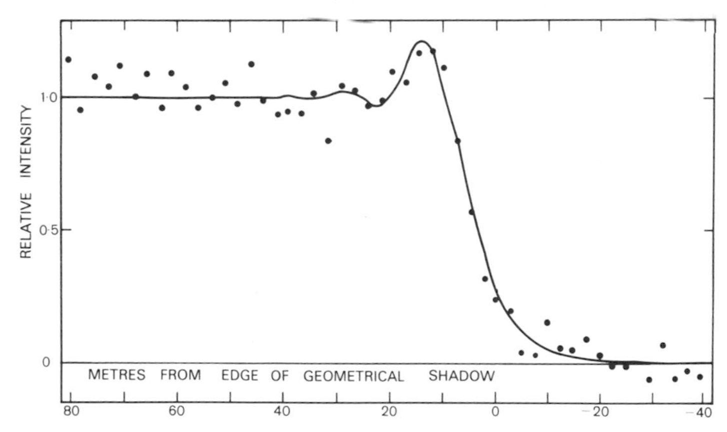

Figure 9-15. Occultation reappearance of Y Tauri. Data is plotted at 4.0 millisecond intervals.

Using the methods outlined in Section IV, a theoretical diffraction pattern for a star of some finite diameter can be constructed, and fitted to the data in the same way as for the point-source case. Now, however, the angular diameter is simply treated as a fifth parameter to be varied. Doing this for Y Tauri yields an angular diameter of 0."0078 ± 0."0013 (solid line, Figure 9-15), assuming a fully limb-darkened disk. This is by no means large as stars go, with R Leonis, the largest measured star, having a diameter of 0".067.

VIII. CONCLUSIONS

This chapter has attempted to provide a thorough introduction to the theory, methods, and uses of photoelectric observations of lunar occultations. The topic remains one of considerable importance, and while participation up until now has been largely restricted to the professional observatories, the next few years will see a steadily increasing involvement by amateurs.

IX. FURTHER READING

Much of the preceding has to date only appeared in the research articles mentioned in the text and listed below. For a good, basic, popular level introduction though, see Evans (1977) and Evans, Barnes, and Lacy (1979). A more technical introduction is the five papers by Nather, Evans, and McCants mentioned in Section I. Further general discussions can be found in "Highlights of Astronomy," Vol. 2, p. 585 (IAU, 1971) and in "The Moon," Vol. 8, p. 490 (1973).

REFERENCES

Africano, J. L.; Cobb, C. L.; Dunham, D. W.; Evans, D. S.; Fekel, F. C., and Vogt, S. S. (1975). Astron. J. 80, 689-697.

Allen, C. W. (1973). Astrophysical Quantities (Athlone Press).

Barnes, T. G.; and Evans, D. S. (1976). Mon.Not.Roy.Ast.Soc. 174, 489-502.

Bevington, P. R. (1969). Data Reduction & Error Analysis for the Physical Sciences (McGraw-Hill).

Dunham, D. W.; Evans, D. S.; McGraw, J. T.; Sandmann, W. H., and Wells, D. C. (1973). Astron. J. 78, 482-490.

Evans, D. S. (1970). Astron. J. 75, 589-599.

________ (1971). Astron. J. 76, 1107-1116.

________ (1977). Sky and Telescope 54, Nos. 3 and 4.

________, Barnes, T. G., and Lacy, C. H. (1979). Sky and Telescope 58, 130-134.

Hall, D. S., and Genet, R. M. (1982). Photoelectric Photometry of Variable Stars (Minuteman Press).

Hewlett Packard (1965). Applications Note No. 52.

Hoffleit, D. (1964). Catalogue of Bright Stars, Yale University Observatory.

Iriarte, B., Johnson, H. L., Mitchell, R. I., and Wisniewski, W. K. (1965). Sky and Telescope 30, 1-11.

Johnson, H. L. (1966). Ann.Rev.Astron.Astrophys. 4, 193-206.

Muller, P., and Meyer, Cl. (1969). Troisieme Catalogue D'Ephemerides D'Etoiles Doubles. (Publ. de l'Obs.de Paris).

Nather, R. E. (1970). Astron. J. 75, 583-588.

________, and Evans, D. S. (1970). Astron. J. 75, 575-582.

________, and McCants, M. M. (1970). Astron. J. 75, 963-968.

10. PORTABLE HIGH SPEED PHOTOMETER PROJECT

Peter C. Chen

I. INTRODUCTION

The Moon, as is well-known, moves against the background of stars. This rather bland statement of fact leads to some rather dramatic events every month with regularity. At the appropriate times, one can see through the telescope a bright star shining steadily in the field while off to one edge the limb of the Moon creeps up to it slowly and ponderously. The Moon comes closer and closer to the star and finally, in a breathtaking instant, the star disappears. If one were patient and kept the telescope tracking for one hour or so, one could see the star suddenly pop up again on the trailing limb of the Moon, apparently none the worse for having been "eaten up" or occulted.

At the instant that the starlight is cut off by the edge of the Moon the light does not suddenly go down to zero. If one were to record the intensity at a rate of one thousand times a second (one millisecond time resolution) one would actually see a series of bright flashes decreasing gradually to zero. This is the phenomenon known as diffraction--a light wave passing close to a sharp edge forms a series of alternately bright and dark zones on the far side. The limb of the Moon forms a sharp edge, and the relative motion of the Moon and the Earth causes the fringe pattern to sweep past a stationary observer. Furthermore, this high-speed record of light change at the instant of occultation can be analyzed to yield a precise position of the lunar limb at the instant of occultation, the measurement of stellar angular diameters in special cases, or the detection of close double stars that are not resolvable by other means. To date, more data on stellar diameters and duplicity of very close systems have been obtained by the occultation method than by any other means.

Photoelectric occultation observations are ideally for amateur astronomers and small college observatories. A moderately large telescope of 8- to 30-inches aperture is needed. A typical observing session consists of a few nights per month. The pace of observing is also enjoyable in that all the disappearances happen in the early evening hours, typically at the rate of one event per hour. On the other hand, the pace quickens as the occultation event approaches, and one must respond instantly at the moment of occultation in order to capture the data. From the scientific point of view, the data from many observers at different sites not only provides insurance against bad weather but also yields the direction and separation in the case of double stars. Moreover, anything unusual can be corroborated. In the case of occultation of stars by planets or asteroids widespread participation is crucial, as these events are very rare and visible only from a very narrow path on Earth which seldom passes through fixed observatories. The scientific payoff is also high as one can, for example, help in delineating the structure of the rings around Uranus or discover the as yet unconfirmed rings or satellites around asteroids.

The chief obstacle to more widespread amateur participation in occultation observations has been the lack of suitable instrumentation. The equipment in use at McDonald Observatory, for example, is simply too expensive and too bulky. What is needed is a simple and inexpensive photometer and computer unit that is reasonably easy to put together and to use. Towards this end, a program had been carried out at the University of Texas under the leadership of Drs. R. Nather and D. S. Evans to develop such an apparatus. This chapter describes the resulting McDonald occultation photometer.

II. A DESCRIPTION OF THE PHOTOMETER

The main criteria of the photometer's design are that it must be small, affordable, reasonably reliable for use in the field, and can be put together and used without requiring special mechanical or electronics skills. It must be able to record photoelectric readings at a rate of up to 1 msec per channel (one thousand times a second), synchronize with a WWV clock to the same degree of accuracy, and provide a real time display of the data. The data can be recorded on ordinary audio cassette tapes or transmitted directly to a personal computer for display and analysis. Such a system is shown in Figure 10-1.

A block diagram of the data system is shown in Figure 10-2. The photometer head unit, which mounts on a telescope in the regular 1 1/4-inch draw tube, contains the usual photometer optics as well as a photomultiplier tube (PMT), a high voltage supply, and pulse amplified electronics. Light from a star is focused by the telescope onto the aperture and either goes up to a viewing eyepiece or through the Fabry lens to an EMI 9781A (improved 1P21 type) photomultiplier. The PMT converts the light into electrical pulses which are then amplified by a LeCroy AA100B pulse amplifier and sent down a coaxial cable to the microcomputer controller unit. The high voltage supply is a single unit DC-DC converter.

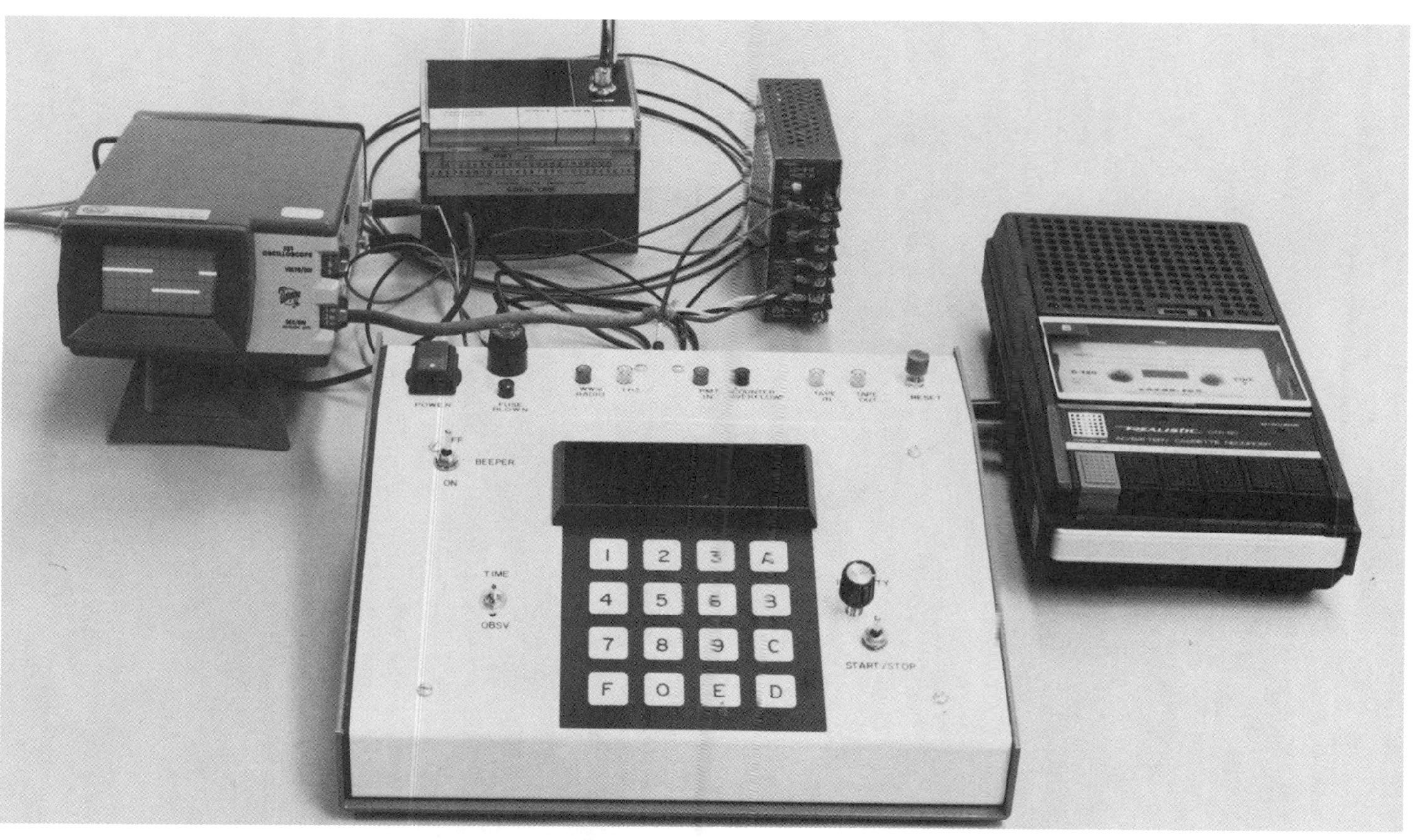

Figure 10-1. The portable occultation recorder.

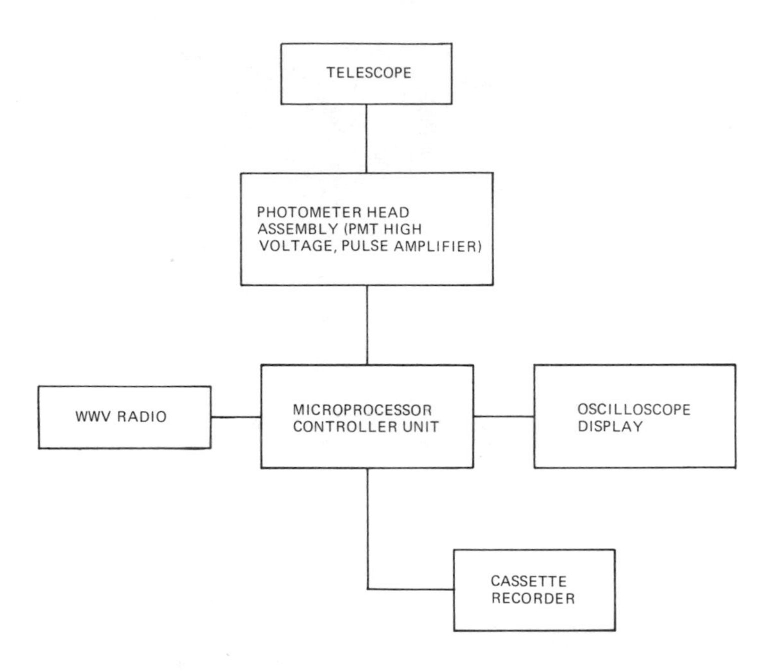

Figure 10-2. Block diagram of portable occultation photometer.

The microprocessor controller unit (see Figure 10-3) functions as follows. At the heart is an Intel 8085 microprocessor CPU (central processing unit) with a 6.144 Mhz crystal clock. The CPU can, under program control, detect the time pulses from a WWV radio that is connected to it via a fiber optic link. This enables the controller to trigger its clock in synchronism with standard UTC time. The photoelectric signal from the photometer unit is fed into a 12-bit CMOS counter which is read at intervals of 1, 2, 4, or 8 milliseconds depending on the choice of the observer. The counts are recorded in an internal memory continuously in a circulating or ring counter mode; i.e., data is stored successively in memory locations 1 through 4096, and then started from 1 again (and thereby erasing the contents of the previous cycle). Current data is displayed on an oscilloscope unit; we are experimenting with a LED flat panel display as well as with a portable black and white television set. When the occultation has occurred as evidenced by the oscilloscope display, the observer commands the controller to stop and thereby freezes a particular set of data. These are then read out to an audio cassette recorder or through a serial output to a host computer. At a later time, a particular observation can be read back into the controller memory for display and detail examination, or for transfer to a host computer in a standardized format.

A special asteroid mode is also included for asteroidal or planetary occultations. The requirements are different in this case because of the very slow angular movement of asteroids in the sky and the very long resultant time scales involved. Typically,

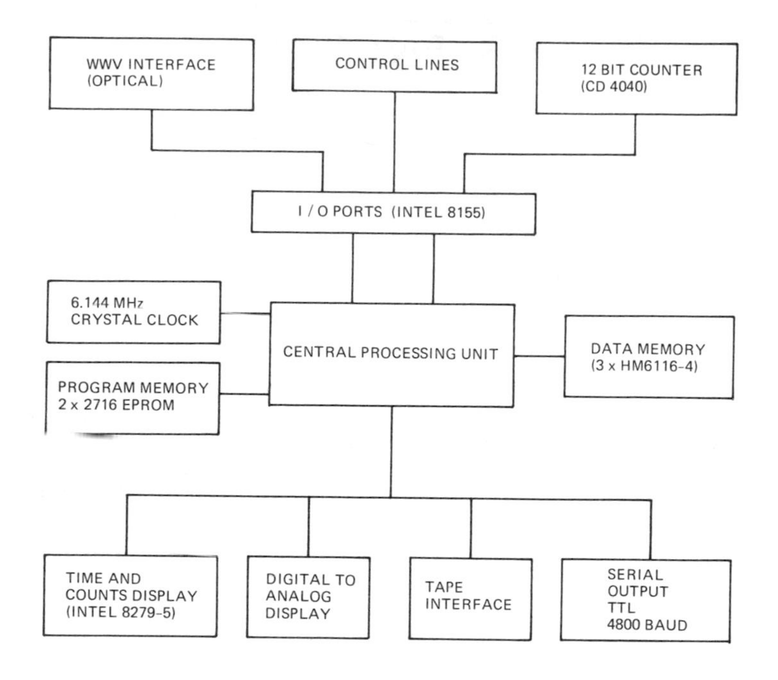

Figure 10-3. The microprocessor controller unit.

one needs only to record photometric data at the rate of 10 times a second for up to an hour; this, however, requires much more memory than is available with our device. We have therefore devised a special operating mode for slow occultations. When in this mode, the controller (once synchronized with WWV) begins to collect data at intervals of 100 msec (0.1 seconds) and outputs the data stream continuously to a tape recorder. A display of the data collected during the previous 30 seconds is also provided on an oscilloscope. The data collection proceeds until either a stop signal is entered by the observer, upon which the controller outputs the time of end of observations and exits, or until all the tape is used up.

A series of special routines have been written to enhance the operation and data transfer abilities of the controller unit. For example, during data acquisition, one can choose between two modes of display. In one mode, the display is stationary on the screen and sweeps continuously through data memory, while in the other mode the newest data point appears at one edge of the screen and moves to disappear at the opposite edge. In both of these modes the display level can be moved up and down or magnified by arbitrary amounts; these operations simulate the adjustments of the baseline level and gain settings that facilitate the detection of occultation events on a DC amplifier photometer. A set of simple instruction is available to check the degree of synchronism between the system clock and the WWV signals, and to correct divergences to the 1 ms accuracy required. Finally, data can be examined point-by-point after an observation or from a previous record on tape.

III. CURRENT STATUS

At the present time, five controller units and two photometer heads have been built. The controller units are being tested by various colleagues and amateur groups in Dallas, Texas; San Antonio, Texas; Pomona, California; Carter Observatory, New Zealand; and Shanghai Observatory, China. Observations of good quality have been reported from New Zealand and from Dallas, the latter using a 14-inch Celestron telescope. The rest are expected to be producing shortly, and plans are being made for a series of coordinated lunar occultation observations to be carried out in early 1983.

We are currently working on an optimization of the system, both to enhance its operation and to simplify its construction. We have expressions from several firms interested in its commercial production. Mr. Arthur Sweeney of the Dallas Astronomical Society is also collaborating with us to convert our occultation data reduction program for use on home computers.

IV. TECHNICAL DETAILS

Size:	Controller - 8 1/2 x 10 1/2 inches (slanted) Photometer head - T shaped, maximum diameter 5 x 5 1/2 x 1 1/2 inches
Weight:	Controller - 4 pounds Photometer head - 3 pounds
Microcompressor:	Intel 8085A, 3.072 MHz (6.144 MHz clock)
Counter:	Opto-isolated, expects external pulse source with maximum frequency = 1 MHz, minimum pulse width 500 ns 12-bit length, maximum count = 4095
Serial output:	TTL, 4800 baud 9 bit ASCII-1 start bit + 7 data + 1 stop bit, no parity 80 column card image format, consisting of 10 data per line plus 4 digit counter
Power requirements:	8 to 10 volts, DC, 0.5-0.7 amps 12 volts with dropping resistor

This equipment has been produced under NSF Grant 81-06063 (David S. Evans, Principal Investigator).

INDEX